Long-Term Perspective in Coastal Zone Development

Frank Ahlhorn

Long-Term Perspective in Coastal Zone Development

Multifunctional Coastal Protection Zones

Frank Ahlhorn
Universität Oldenburg
Zentrum für Umwelt- und
Nachhaltigkeitsforschung
(COAST)
26111 Oldenburg
Germany
frank.ahlhorn@ewetel.net

ISBN 978-3-642-01773-5 e-ISBN 978-3-642-01774-2
DOI 10.1007/978-3-642-01774-2
Springer Dordrecht Heidelberg London New York

Library of Congress Control Number: 2009926866

Cover design: deblik, Berlin

Photo of the Author: (c) Nordwestzeitung, Red. Varel.

Printed on acid-free paper

Springer is part of Springer Science+Business Media (www.springer.com)

We need plans for the future
that do not paint all grey and black,
but formulate worthwhile aims.

Source: Prof. Dr. Hans-Peter Dürr
(Recipient of the Right Livelihood Award in 1987)

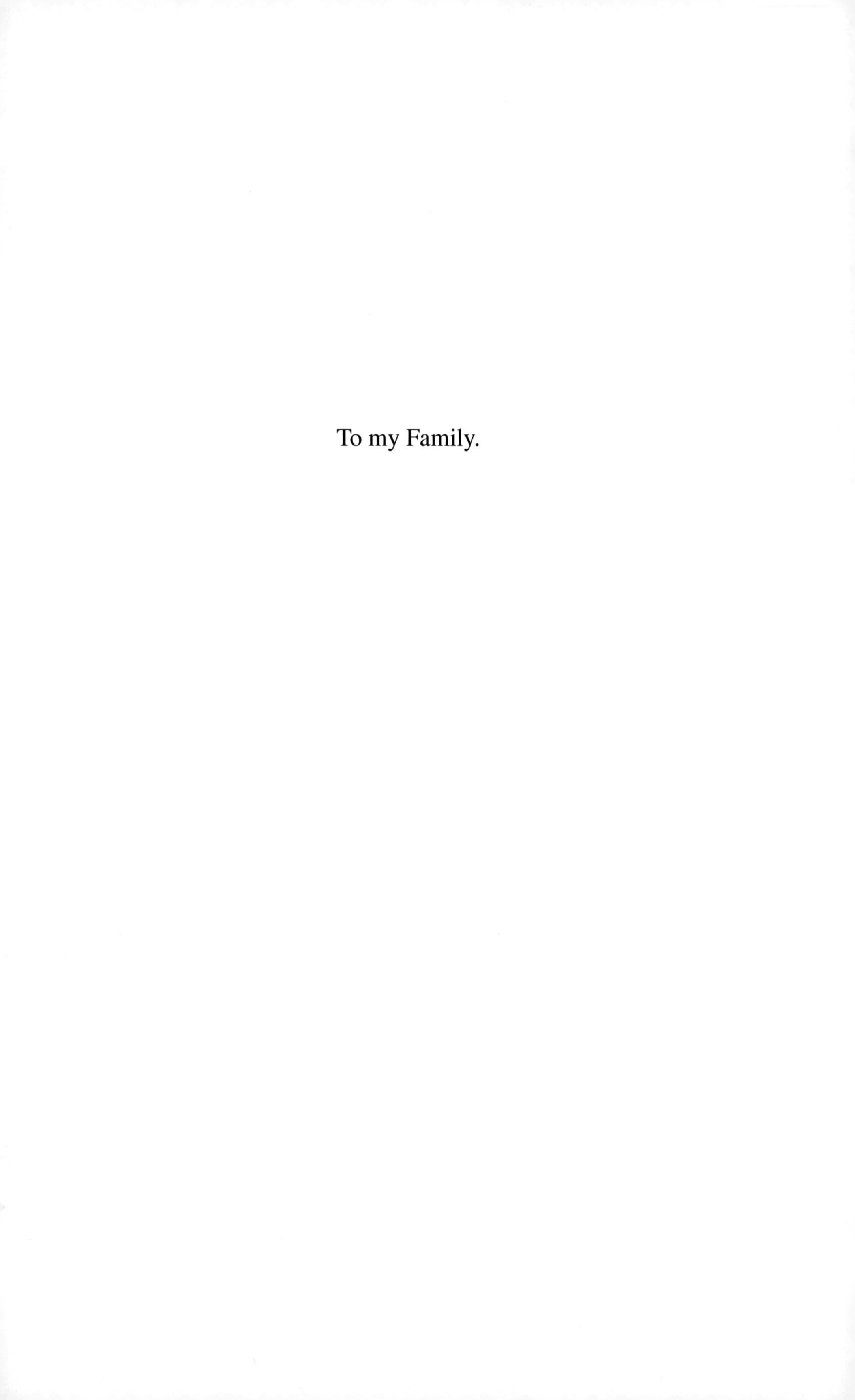
To my Family.

Preface

From February 2004 until December 2007 the ComCoast project (*Com*bined Functions in *Coast*al Defence Zones) was carried out within the Interreg IIIB framework. The objective of this project was to investigate and to test the options for spatial coastal protection concepts. The mission statement of the project over its entire duration was: *A wider approach in coastal thinking*. I was glad to be part of this project and would like to thank all my European colleagues for the constructive and cooperative atmosphere.

Firstly, I would like to express my thanks to Prof. Dr. Horst Sterr and Prof. Dr. Hans Kunz who assisted me over these years with their thorough guidance and comprehensive expertise and for many in-depth discussions on the focus of this dissertation.

Secondly, I would like to thank my colleague Dr. Jürgen Meyerdirks, who acted as my key scientific sparring partner over the last two years, and who continuously demonstrated his ability to bring me back to earth. Additionally, I would like to thank my colleague Dr. Thomas Klenke, who gave me the chance to be part of a challenging European project consortium and who supported me with his expertise on the project and the dissertation. Furthermore, I would like to express my special thanks to Gerard McGovern, who thoroughly improved my English.

Last, but not least, I owe my special thanks to my family – Malika, Marten, Jonte, Lasse and my wife Sandra – who supported me with patience, creative interruptions and forced me to finally complete my dissertation.

Contents

List of Figures

List of Tables

Chapter 1
Introduction

Contents

1.1 Aim, Context and Guiding Questions

The aim of this dissertation is to develop and to discuss an approach towards a sustainable use of the coastal zone addressing the dual issues of safety and development. This approach consists of the concept of Multifunctional Coastal Protection Zones (MCPZ). A MCPZ is an area where the application of spatial coastal protection concepts offers new options for safety and development for society, ecology and economy. The options of a MCPZ will be elaborated within the framework of a Participatory Integrated Assessment (PIA) process. The focus area is the southern North Sea Region. This dissertation will concentrate on the German sector, especially on Lower Saxony.

The following section serves to provide a general overview of the international and national context of the contribution this dissertation seeks to achieve: the starting point is the awareness of human interference in the global climate and the work of the World Commission on Environment and Development (WCED) in 1987 and the installation of the Intergovernmental Panel on Climate Change (IPCC) by the United Nations in 1988. Since 1992, after the Earth Summit in Rio de Janeiro (UN 1992), the concept of sustainable development has been established and placed on the political agenda. Triggered by and accompanied by the awareness of human interference in the global climate (IPCC 1990), the concept of sustainability is cited in a growing number of laws, plans and regulations. Within this dissertation the concept of sustainable development will be understood as a holistic approach to consider all relevant functions, resources and types of land use in coastal zones in line with the definition given by the WCED (1987): *Development that meets the needs of the present generation without compromising the ability of future generations to meet their own needs.*

F. Ahlhorn, *Long-Term Perspective in Coastal Zone Development*,
DOI 10.1007/978-3-642-01774-2_1,

IPCC has different groups with several subgroups focussing on specific issues. Working Group III on Coastal Zone Management is charged with providing information and recommendations to national and international policy on coastal zone management strategies and long-term policies on adapting to climate change and sea level rise. At the Coastal Summit (World Coast Conference 1993 in Noordwijk, The Netherlands) scientists and politicians from approx. 100 countries discussed solutions for the challenges of climate change and sea level rise (Bijlsma 1994). In 1992, the German coastal scientific community together with the association *Eurocoast* (installed in 1989 by the European Commission as a working group of coastal experts) organised a coastal forum to discuss a wide range of issues linked to the challenge caused by climate change, sea level rise and multiple uses at the coastal zone (Sterr et al. 1992). Simultaneously, Germany launched the research programme "Climate Change and the Coast" (Schellnhuber and Sterr 1993) dealing with climate change and its consequences for the German coastal zone. The European Commission has stressed an integrated management of coastal zones since 1995 (EC 1995). In the following years Integrated Coastal Zone Management (ICZM) has been established as a research as well as a policy field to meet the challenges of climate change and future land use. The EU has launched a demonstration programme to develop common principles and to enhance the commitment to an ICZM. The results and experiences of this demonstration programme were the basis for the EU recommendations on ICZM (EC 2002). Sustainable development addressing both safety and development issues is crucial for the development of low lying coastal areas. The necessary precondition for social and economic development is safety against flooding caused by both storm surges and/or high fluvial water. Thus, coastal protection has to be integrated into a sustainable development strategy for the coastal zone, see e.g. Kunz (1991), Hillen et al. (1992), Ahlhorn and Klenke (2006b), Klenke et al. (2006).

Combining both, sustainable development and integrated coastal zone management, integrated sustainable development of coastal zones is the challenge to be faced. To meet this challenge, it is necessary to consider all relevant aspects that pertain to coastal development. In coastal zones, coastal protection is a necessary precondition for safe living and working. Consequently, coastal development schemes (plans) have to consider a comprehensive matrix of interests, needs and functions. The interdependencies are evident: land use influences coastal protection concepts and vice versa. These interdependencies will be elaborated in more detail within this dissertation and the challenges of climate change and possible reactions will be shown in the treatment of the MCPZ concept.

This short overview demonstrates the necessity to deal with the future challenges caused by climate change and the influence of instruments developed at international and European levels. Considering the slogan: *think global, act local* the global situation has been roughly described above. With respect to the German situation, the following reasons are asking for solutions over the next decades: adaptation of the traditional single-line embankments to a rising sea level causes problems (now and in the future); heightening the main dike demands more resources (material, space and money). These problems and the changing attitudes towards a sustainable (coastal) development over the last decades call for new approaches. Additionally, the likely effects of climate change on the coast have to be taken into account. The uncertainty inherent in these effects has to be treated adequately, thus, the attitude of the coastal protection authority to decide on the basis of secure knowledge should be changed. The risk management approach offers a framework to deal with these uncertainties, and tables an agenda comprising e.g. risk assessment, risk perception, risk communication.

This dissertation will concentrate on the concept of MCPZ which tries to offer new options. It was elaborated within the framework of the EU Interreg IIIB project *ComCoast* (*Com*bined functions in *Coast*al Zones). Ten institutions from five countries around the southern North Sea collaborated in this project. The motto of the project was: *A wider approach in coastal thinking*. *ComCoast* developed and demonstrated innovative flood risk management and coastal protection strategies to stimulate wider functions along the coast and a more gradual transition from sea to land (RWS-DWW 2006).

Consequently, the guiding questions for this dissertation are as follows:

1. What are the major drivers today for new approaches in the future for coastal protection and spatial planning?
2. Are the existing instruments and methods of coastal protection and spatial planning capable of adapting to the new challenges?
3. Which options do spatial coastal protection concepts, especially MCPZ, offer and how can they be implemented?

1.2 Structure of the Dissertation

This dissertation is divided into four parts: (a) description of the natural and social background, (b) elucidation of the challenges concerning the future of both natural and social development, (c) description of the existing instruments and tools for the management of the participation process and their suitability for implementing options, resolving gaps and bottle-necks detected during the implementation of MCPZ and finally (d) the discussion of the implementation of MCPZ by a PIA process (see Fig. 1.1).

In Chap. 2 the natural environment and the social background of coastal protection will be explained. Throughout human history people have settled at the coast and made use of the existing natural resources. Without coping with the specific natural conditions the natural potential of the coastal zone functions like agriculture, tourism, shipping, etc. could not be exploited. Hence, the natural conditions will be described shortly in the first part of Chap. 2. Relevant information will be explained to classify the interests and needs of several types of land use in the coastal zone, e.g. nature conservation, tourism/recreation, agriculture and coastal protection. Settlers

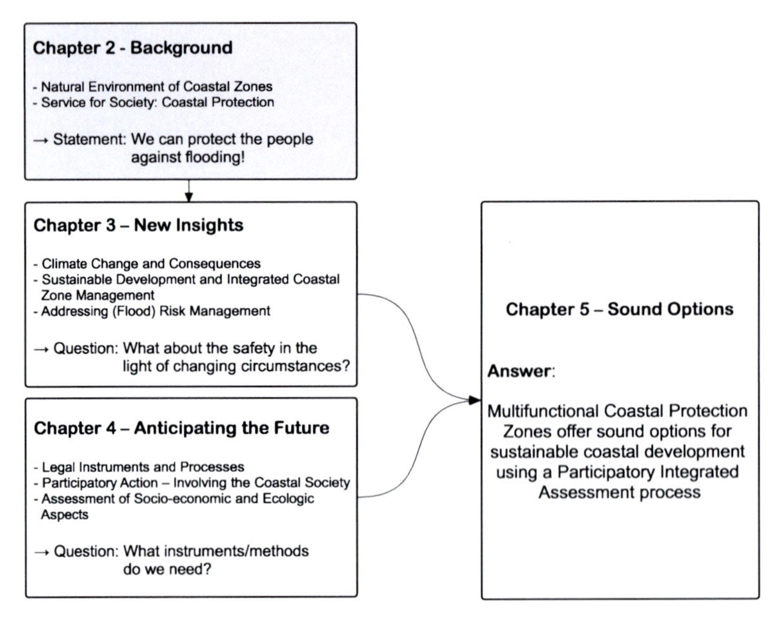

Fig. 1.1 Structure of the dissertation

have always had to face the risks inherent in coastal zones, e.g. storm surges and high fluvial water from the hinterland. In order to understand today's scenarios, it is necessary to review some key historical decisions and the development of society. The description of historical and existing coastal protection concepts of two European countries will close this chapter.

The basis for a new awareness of the need for sustainable coastal development (Chap. 3) are the probable consequences of climate change. The IPCC was established in 1988 and the first assessment report was published in 1990. On international level the IPCC was the first stimulation to think about redirecting development policy, especially on climatic aspects. The next step was the Earth Summit in Rio de Janeiro in 1992. The process of sustainable development initiated in Rio will briefly be described as a forerunner of ICZM, which was highlighted in Chap. 17 of the Rio Declaration. Society is confronted with two reaction strategies: *adaptation* and *mitigation* (IPCC 1996). Several instruments, approaches and methods have been developed to meet these challenges. Furthermore, dealing with coastal zones, some countries have moved from coastal protection towards flood risk management. That means not only fighting against the sea but dealing with the risk situation in a continuous and holistic way. Methods and concepts used in risk assessment and risk management will be discussed in Sect. 3.3. Section 3.4 addresses the collation of unresolved problems of coastal zones along the southern North Sea, especially in Germany and in The Netherlands. The changes mentioned, e.g. climatic, will impose new challenges and unresolved problems will generate new demands on safety and development.

Chapter 4 deals with the capability of instruments and methods to anticipate and adequately react to the challenges explained and elaborated in Chap. 3. The introduction of ICZM and the concept of sustainable development, especially sustainable coastal development, demands new and innovative methods e.g. for participation and evaluation. For the implementation of spatial coastal protection concepts it is necessary to involve and engage stakeholders early in decision-making processes. Methods and instruments of public participation will be discussed in more detail in Sect. 4.2. Socio-economic evaluation – the costs of different coastal protection projects in combination with different kinds of land use, taking social, ecological and economical aspects into account – is the focus of Sect. 4.3. Chapter 4 closes with the elaboration and discussion of the implementation framework for the Participatory Integrated Assessment (PIA) process deduced from the experiences of the ComCoast pilot projects.

In Chap. 5 the territorial definition of a coastal protection zone on the Lower Saxonian coast will be described and necessary actions will be elaborated and discussed in detail. A brief description of spatial coastal protection concepts proposed by the EU Interreg project ComCoast is provided in Sect. 5.1. Afterwards, the entire process of the PIA applied at the German pilot area Nessmersiel will be explained: the procedure for identifying feasible implementation sites, the design and setting of the participation process, the applied methods and tools and the discussion of the process results. Chapter 5 ends with the lessons learned and a discussion of the options for future projects disclosed by the PIA process.

In Chap. 6 the guiding questions will be answered and recommendations given.

The dissertation closes with summaries in English and German.

1.3 Methodological Approaches

For the purpose of this dissertation, many research fields have had to be combined and their interdependencies evaluated such as participation, coastal protection, climate change, risk management, socio-economic evaluation. The chosen methodological approach to achieve the goal for the evaluation whether the application of MCPZ is possible or not, will be briefly explained.

The dissertation is based on both existing methods and the development of new approaches to implement MCPZ. Established methods were used and adapted to the special situation and the requirements of the PIA process. The applied and created methods and tools are as follows:

Participatory Integrated Assessment (PIA): The participation process for the German pilot area was structured as a Participatory Integrated Assessment process. The process was divided into four steps: kick-off meeting, planning exercise with scenario-technique, consensus workshop about the weighting of criteria and the participatory assessment. The approach of PIA has already been applied to several problems in water management and industrial companies, but the application of this process in coastal zones is quite new.

Planning Exercise with Scenario Technique: This method was applied at the German pilot area Nessmersiel. Assumptions about climate change and the consequences provided the basic settings presented to the participants together with information on coastal protection. The results of the planning exercise were three integrated scenarios which serve as input data and information for the following steps of the PIA process.

Design Elements (DE): The Design Elements were introduced to bridge the gap between local and regional planning, i.e. as concrete as necessary. The DE's are assigned to a certain value category (use value, functional value or existence value) and therefore have to fulfil other requirements to be used as spatial criteria within the participation and within the evaluation process. The DE's were implemented within the GIS to visualise the maps and provide a communication platform. They were also instrumental in determining the characteristics within each integrated scenario.

Geographic Information System (GIS): A Geographic Information System is a well-known and widely applied tool which was used within several parts of the project. The advantage and purpose of a GIS is the ability to visualise and to analyse spatial information. A GIS was applied to identify feasible areas for spatial coastal protection concepts. Furthermore, the GIS was used for the socio-ecological-economical evaluation.

Socio-economic-ecologic Evaluation: Cost-Benefit-Analysis serves to enable socio-economic evaluations and comprises a range of methods. These however, display certain disadvantages in that they require reliable figures to calculate the best cost-benefit ratio. The application of a scenario-technique and the integration of several criteria for socio-economic and ecologic aspects demand a specific evaluation method. An outranking method was chosen which best met the requirements of multi-criteria decision aid. This method can handle quantative as well as qualitative data and information, which is crucial, dealing with different value categories like functional value or existence value.

Structured Interview: This method was applied to gain information and feedback. One-to-one interviews were conducted to obtain feedback from members of the participation process of the German pilot area and on the products of the ComCoast project. The interviews were used to obtain information on the characteristics of the DE's for the evaluation method.

Chapter 2
Background Information – Nature and Society

Contents

2.1 The Natural Environment of the North Sea Coastal Zones

2.1.1 Basic Information

The North Sea is a shallow sea adjacent to the northern Atlantic. The southern North Sea is divided into two parts: the Southern Bight and the German Bight. The mean depth of the southern North Sea is approx. 20–30 m. The recent shape of the North Sea is a result of fluvial and fluvio-glacial processes during and after the ice ages, Sindowski (1962), Streif (1982, 2002), Behre (2007).

The North Sea has a large variety of landscapes along its coast: e.g. cliffs, firths, Wadden Sea, dune areas, and fjords. The East Coast of England is characterised by estuaries such as Humber and Thames, and by further expanses of sand and mud flats in areas such as The Wash. Along the Channel the coastline of south-east England is dominated by low cliffs and flooded river valleys. From East to West along the French coast of the Channel the North Sea offers maritime plains and estuaries, cliffs, and the rocky shore of Brittany.

F. Ahlhorn, *Long-Term Perspective in Coastal Zone Development*,
DOI 10.1007/978-3-642-01774-2_2,

From the Strait of Dover to the Danish West Coast, sandy beaches and dunes prevail with numerous estuaries (e.g. Scheldt, Rhine, Meuse, Weser and Elbe) and the islands of the Wadden Sea with their tidal inlets. In Denmark large lagoon-like areas exist behind long sandy beaches.

In main figures: the coastline is about 36,000 km long, the land-area within the 10 km zone is approx. 127,500 km^2, the population of this area is approx. 165 million. In comparison with other European coastal areas the North Sea has the highest level of urbanisation (17% of the coastal zone), the highest armouring of the coast including defences and harbours 20% of the North Sea coast is eroding and the highest level of protection in terms of the number of NATURA 2000 sites (EEA 2006, p. 20).

The following paragraph provides a short overview of the hydrography and meteorology of the North Sea with a focus on its southern part. The hydrographical and meteorological conditions are the main forces which create the great variety of landscapes along the North Sea coasts.

2.1.1.1 Hydrography and Meteorology

The different landscapes along the North Sea coast coincide with the existing tidal range (Fig. 2.1). Due to the long connecting line with the northern Atlantic the North Sea has significant tidal waves. Without this connection, there would only be micro tidal waves as in the Baltic Sea. The tidal range between Den Helder (The Netherlands) and Borkum (most western barrier island of the German Wadden Sea) increases from about 1.5 m up to 2.3 m (high meso-tidal). Further to the East (Inner German Bight) the tidal range increases up to 3.6 m near Wilhelmshaven (Jade Bay) and to approx. 4.2 m in the city of Bremen 70 km downstream the river Weser (low macro-tidal) (Niemeyer and Kaiser 1999).

The duration of a tide is 12 h and 25 min (semi-diurnal period). In general, the tidal wave in the Wadden Sea is influenced by the topography of the area, by the planet constellation, by meteorological conditions, by the amount of fresh water discharge of the rivers (Niemeyer and Kaiser 1999). "Tidal currents are the most energetic feature in the North Sea, stirring the entire water column in most of the southern North Sea and the Channel. In addition to its predominant oscillatory nature, this cyclonic propagation of tidal energy from the ocean also forces a net residual circulation in the same direction" (OSPAR 2000, p. 19).

The North Sea is situated in the temperate zone with a climate mainly influenced by the inflow of oceanic water from the northern Atlantic and by the large scale westerly air circulation which frequently contains low pressure systems. The climate development of the North Sea is directly linked to the large scale atmospheric circulation in the European-Atlantic system. The North Atlantic Oscillation index (NAO index) indicates the influence of high pressure at the Azores and of low pressure in the North of the Atlantic: a higher NAO index indicates the generation of a stronger west wind circulation and in consequence, stronger westerly winds creating higher water levels in the North Sea (Weisse and Rosenthal 2002). The strength

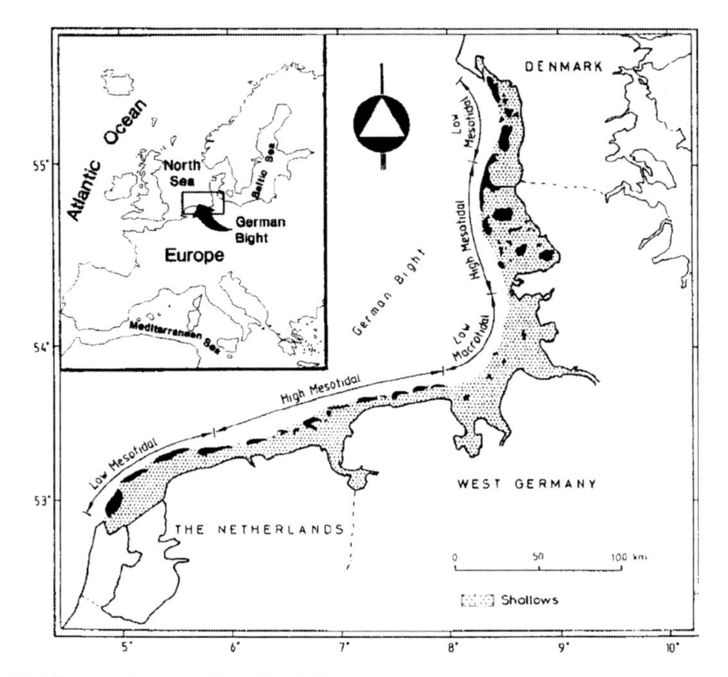

Fig. 2.1 Tidal range of the southern North Sea
Source: Dieckmann (1992).

of the westerly winds has a significant effect on water transport and distribution, vertical mixing and surface heat flux. This "atmospheric circulation" is also closely related to the cloud cover and therefore the light conditions in the water column and the coastal zones (OSPAR 2000). Moreover, other climatic-oceanographic features related to the NAO index include: temperature, salinity and circulation. A stronger NAO index causes a stronger influence of the northern Atlantic correlated with an increase of material transport, higher salinity and an increase of temperature (Weisse and Rosenthal 2002).

Finally, "the North Sea climate is characterised by large variations in wind circulation and speed, a high level of cloud cover, and relatively high precipitation. Rainfall data show precipitation ranging between 340 and 500 mm per year, and averaging 425 mm per year" (OSPAR 2000, p. 22).

A short description of typical landscapes in the southern North Sea Region is shown in Table 2.1. Selected landscapes will be described later in more detail providing a glimpse of the natural environment of the case study areas in Germany and the ComCoast partner countries.

2.1.2 Landscapes and Important Areas

The southern North Sea region is dominated by two landscapes: Dune Areas (mainland and barrier islands) and the Wadden Sea. They are the natural starting point

Table 2.1 Description of different landscapes along the southern North Sea

Elbe and Weser estuary, Jade Bay	The Elbe and the Weser discharge through their estuaries huge volumes of (contaminated) fresh water into south-eastern corner of the North Sea and into the Wadden Sea.The Jade Bay is a Wadden Sea-like tidal inshore basin connected to the open sea by a narrow channel. All three have important shipping lanes and are thus subject to intensive dredging and deepening. The Elbe and Weser have a strong and vertical salinity stratification although tidal and wave activity can be very strong. In the Jade Bay small fresh water input and very strong tidal currents suppress the development of stratification
Wadden Sea (including Ems-Dollart)	The Wadden Sea extends along the North Sea coasts of The Netherlands, Germany and Denmark, from Den Helder to the Skallingen peninsula near Esbjerg. It is a highly dynamic area of great ecological significance. With 500 km it is the largest unbroken stretch of mudflats in the world. According to the delimitation of the trilateral cooperation, the Wadden Sea covers about 13,000 km^2, including some 1,000 km^2 islands, 350 km^2 salt marshes, 8,000 km^2 tidal areas (sub-tidal and inter-tidal flats) and some 3,000 km^2 of offshore areas. Most parts of the Wadden Sea are sheltered by barrier islands and contain smaller or wider area of intertidal flats. During each high tide an average of 15 km^3 of North Sea waters enters the Wadden Sea, thereby doubling the volume from 15 to about 30 km^3. With the North Sea water also nutrients and suspended matter reach the Wadden Sea. In the North of Holland there is also a structural loss of sand from the Wadden Sea. There is a structural loss of sand from the offshore area to the tidal area causing erosion of the foreshore and beaches of several islands
Dutch coastal zone	The coastal zone along the entire western and northern half of The Netherlands can be considered as one of the most densely populated areas in Europe. The coastal zone is protected from the sea by natural sand-dunes (254 km) and sea dikes (34 km), beach flats (38 km) and 27 km of boulevard, beach walls and the like. The width of the coastal dunes varies between less than 200 m, and more than 6 km. The upper shore-face is a multi-barred system generated by normal wave action, while its lower part is dominated by storm sedimentation, down to the depth of about 16 m. At greater depths tidal currents play a significant role along with storm waves, keeping fine-grained sediment in suspension
Scheldt estuary	The Scheldt estuary is well-mixed with a yearly average upstream freshwater flow rate of 107 m^3/s. The total drainage area is 20,300 km^2. The estuary consists of an alteration of transition zones: deep ebb and flood channels, large shallow water zones, tidal flats and dry shoals

Source: OSPAR (2000, pp. 8–9).

for (sustainable) development in the coastal zone. The focus area of this dissertation in Germany is the mainland of Lower Saxony. The Wadden Sea of Lower Saxony consists of dunes on barrier islands, of estuaries, the sheltered (behind the islands) and the open Wadden Sea (without islands) – see Fig. 2.1. The partner countries of the ComCoast project have similar landscapes with slightly different conditions.

The Dutch Wadden Sea extends from the island Texel to the island Rottumer Oog, adjacent to the German border (the Dollart belongs to both countries). In Denmark only a small stretch is covered by the Wadden Sea with the islands of Rømø, Fanø and Mandø. North of the Danish Wadden Sea area there are dunes and sandy beaches.

Between Den Helder (NL) and the estuaries of the Rhine, Meuse and the Scheldt a long sandy coastline with dunes and beaches presents itself. In some places the chain of dunes is disconnected due to storm surge events in former times (e.g. in the proximity of the village Petten, the Hondsbossche and the Pettemer Sea Defence with a main dike between the dunes, see Sect. 3.4 on p. 71. In respect of England the dissertation concentrates on the region of Essex and Suffolk in East Anglia, because the pilot regions of the ComCoast project are located here. The region is dominated by several rivers and their estuaries with many salt marshes.

2.1.2.1 Dune Areas

In the southern North Sea Region dune areas extend on the West coast of The Netherlands between the Delta area (Hoek van Holland in the South) and Den Helder in the North. The coastline of The Netherlands is approx. 350 km long and approx. 250 km are dunes (Hillen and de Haan 1993). The dunes cover about 400 km^2 which is nearly 1% of the Dutch surface (Louisse and van der Meulen 1991). In the Delta area the islands have dunes at their seaward tips which are sometimes very narrow. The northern part, the Holland coast, consists of broader dunes with a length of up to 3.5 km and heights of up to 50 m. The shoreface consists of a breaker zone between Mean Sea Level (MSL) and 8 m depth line (Louisse and van der Meulen 1991). The Wadden Sea area of The Netherlands also features dunes. They cover an area of approx. 11,300 ha and are mainly located on the barrier islands (Petersen and Lammerts 2005).

All barrier islands of the Lower Saxonian Wadden Sea are covered by dunes, and in Schleswig-Holstein this applies to the islands of Sylt, Amrum and Föhr. The barrier islands are formed and sustained by the combined action of wind, waves and tides. Normally, a barrier island consists of a shoreface, beach, dunes and overwash areas. On the mainland side of some barrier islands salt marshes (polders) can be found. The dune area in Lower Saxony covers approx. 4,400 ha and in Schleswig-Holstein approx. 1,500 ha (Petersen and Lammerts 2005). In England sand dunes are rare and widely scattered, but with concentrations along the Lincolnshire and Humberside coasts and in North Norfolk between The Wash and Cromer. The dune area along the North Sea is about 25,000 ha (Doody et al. 1993).

"Dune formation occurs where a supply of dry, wind-blown sand is trapped by an obstacle such as shingle ridge, tidal litter or vegetation. This process often take place above a sand flat which is exposed sufficiently at low tide for the surface layer of sand to dry out. The dunes of the North Sea coast are characterised by the creation of front shore sand ridges formed by the opposing forces of prevailing and dominant winds which occur as offshore and onshore winds, respectively" (Doody et al. 1993, pp. 7–8). Shingle fringing beaches are highly mobile and may not support vegetation communities. Stable and semi-stable vegetated shingles are concentrated in Shetland, Orkneys and East Anglia. They can mainly be found in the south-east, from Norfolk to East Sussex. Altogether the shingle area is approx. about 2,750 ha (Doody et al. 1993).

2.1.2.2 Wadden Sea

The Wadden Sea area is divided into a Dutch, Danish and a German part. The seaward border is the 12-nautical-mile-zone and landwards the main dike line. The mainland adjacent to the Wadden Sea provides a living and working environment for approx. 3.3 million inhabitants (WSF 2005). The following section describes in brief important features and elements of the Wadden Sea from the perspective of the dissertation objectives. More detailed and comprehensive descriptions of the Wadden Sea can be found for example in Abrahamse et al. (1976), Reineck (1978), Ehlers (1988), Buchwald (1991), Lozán et al. (1994), Gätje and Reise (1998), NLP-V and UBA (1999), TERRAMARE (2001), Essink et al. (2005).

The Wadden Sea area is subject to tidal influence and therefore it is classified in several tidal areas (see e.g. Fig. 2.4):

Sub-Littoral: The area below the low water line. The sub-tidal area is divided in an upper and a lower sub-tidal area. The upper sub-tidal area covers the shallow sea in front of the barrier island, and the lower sub-tidal area includes the bigger tidal channels and tidal ebb deltas and is always covered by water. The environment above the water line is the living space for birds and seals.

Eu-Littoral: This area is flooded twice a day and includes the tidal flats and the shore-face. It mainly consists of flats with gentle slopes from the high to the low water line. The tidal flats are flooded and drained by numerous channels.

Supra-Littoral: This area is above the Mean High Tide Water (MThw) and is only flooded at very high water levels. The salt marshes of the mainland and in the mainland side of the barrier islands are a distinctive feature of this area. The salt marshes are carpeted with vegetation, mainly halophytes and in higher regions with salt tolerant plants.

Epi-Littoral: This area contains the dunes on the barrier islands and the area between the embankments and the pleistocene hinterland (until NN +10 m contour line in Germany).

The Wadden Sea area itself is also divided into an outer and an inner Wadden Sea area. The outer Wadden Sea area lies between the water bodies of high and lower salinity in front of the barrier islands. The inner area stretches between the barrier islands and the mainland, containing tidal flats, sand flats, channels and salt marshes. The Wadden Sea is a highly dynamic system with an energy input from the sun, wind, tides and waves. This highly dynamic system underlies natural changes through strong ice winters (e.g. risk for mussel beds), erosion and parasites (e.g. reduction of seals). The German Wadden Sea could be divided into three parts: the North-Frisian part, the East-Frisian part and the Inner Part with open tidal flats.

The North-Frisian part is about 40 km wide and ranges from the islands of Sylt down to the Eiderstedt peninsula. Four types of islands can be found in this area: islands with pleistocene core, marsh islands, Halligen and bigger sand flats. The open Wadden Sea lies between the Eiderstedt peninsula and the river Jade. Within this area the rivers Eider, Elbe, Weser and Jade discard into the Wadden Sea. The tidal flats mainly consist of sand flats and have a gently falling surface. The East-Frisian Wadden Sea stretches from the Jade to the river Ems in the western part of Germany to the border of The Netherlands. The barrier islands in front of the mainland are mainly dune islands, some of them have an older core from pleistocene ages. The Wadden Sea is 10 km wide and consists of 35% of mud and mixed falts. In this area the remnants of older bays can be found: the Jade Bay, the Ley Bay and the Dollart.

2.1.2.3 Tidal Area

The barrier islands are separated by tidal inlets. Tidal inlets are the mouths between the islands where the sediment transport is effected by tidal waters. Within each tidal cycle the water body goes in and out through these tidal inlets and fills and drains the tidal basin between the barrier islands and the mainland. A dynamic equilibrium exists between the tidal currents and the cross-sectional area of the inlet channel (Ehlers 1988, CPSL 2001). "The sediment that is transported by ebb-tidal currents is deposited at the seaward outlet, caused by decreasing current velocities. In result, an ebb-tidal delta develops. However, the erosive forces of deep water waves coming from the North Sea, limit the sediment volume of the deltas. A dynamic equilibrium exists between these erosive forces and the tidal accumulation (Ehlers 1988, Oost 1995, Hofstede 1999). Because the tidal channels of the inlet and the delta are strongly interrelated, they are normally treated as one element" (CPSL 2001, p. 16). On tidal flats the material may become settled as a result of decreasing current velocities. Because the (energy-rich) waves from the North Sea are almost completely dissipated at the shoreface and ebb-tidal deltas (Niemeyer 1986), only local (storm) waves limit the tidal accumulation on the tidal flats. "Similar to the ebb-tidal delta, a dynamic equilibrium seems to exist on tidal flats between the erosive forces of storm waves and tidal accumulation (mainly controlled by the time of tidal inundation)" (CPSL 2001, p. 16).

2.1.2.4 Bays

The Wadden Sea area features a range of bays. Many of them have been reclaimed over the last centuries, like Lauwersoog in the province of Groningen (NL) or the Harle Bay in the north-western part of Lower Saxony in Germany. Existing bays in the Wadden Sea from Lower Saxony to Schleswig-Holstein are the following (from West to North): Dollart, Ley Bay, Jade Bay, Meldorfer Bay and Tümlauer Bay. The historical development of most bays around the Wadden Sea is similar. Severe storm surges in the middle ages caused their largest extension. Afterwards, due to the natural processes of sedimentation and the increasing ability of the coastal community to protect themselves against flooding, land was reclaimed step by step over the years. To illustrate the historical development of German bays the box below (p. 14) contains an extract of the development of the Ley Bay.

Ley Bay – Part I

The Ley Bay witnessed its largest extension approx. 600 years ago, as a consequence of severe storm tides in the middle ages. Until the middle of the last century land reclamation works were executed to increase the arable area for the inhabitants. In the 1950s the Ley Bay was mainly shaped by economic drivers (Erchinger 1970, Hartung 1983, Janssen 1992, Kunz 1999b). Figure 2.2 shows the historical development of the Ley Bay. Approximately 10,000 ha were reclaimed and this new land was offered to inhabitants and refugees of the second world war. After the 1950s the effort to reclaim land from the sea decreased, because of the diversification of working fields after the second world war and the increase of the effectiveness in agriculture. On the other hand, the problems with water management in the hinterland around the Ley Bay intensified, the existing tidal channels silted up continuously. Until 1985 the problems also affected the harbours around the Ley Bay which were dependent on free access to the North Sea. Siltation had imposed increasing pressure on water management and shipping, resulting in the installation of pumping stations with continuously increasing performance (Janssen 1992).

2.1.2.5 Salt Marshes – Extension, Morphology and Ecology

In general, the largest coherent salt marsh area of the world can be found in the southern North Sea region within the Wadden Sea area from Den Helder (NL) to Blåvands Huk (DK). For an overview of salt marshes in Europe see e.g. Dijkema (1987). Salt marshes will be explained in more detail, because they are important elements within the Wadden Sea in respect to both nature conservation and coastal protection as well as for other types of land use (multifunctional use). Salt marshes are the transition zone between sea and land and they fulfil several functions.

Salt marshes exist along most of the shallow coastal waters where marine sedimentation and erosion are balanced. The west coast of England features approx. 22,300 ha of salt marshes the equivalent of nearly half of the total amount of salt marshes in England. "The largest areas of salt marshes in [England] are concentrated around the Greater Thames estuary in Essex and Kent [...]" (Doody et al. 1993, p. 6). The total salt marsh area of the Wadden Sea is approx. 39,000 ha (Essink et al. 2005). Detailed information and data for each country around the Wadden Sea is shown in Fig. 2.3. The latest figures for Schleswig-Holstein in Stock et al. (2005) display a total area of 11,625 ha.

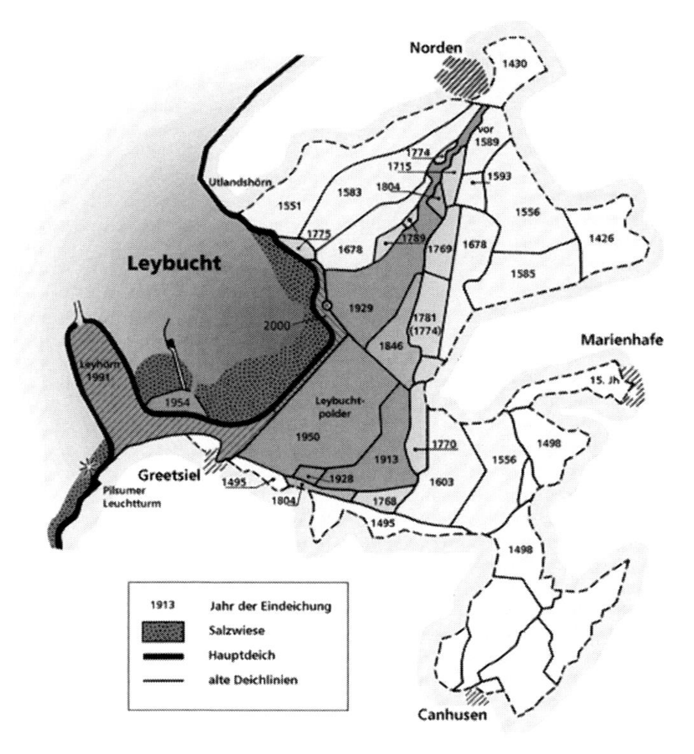

Fig. 2.2 Land reclamation works in the Ley Bay adapted after Homeier (1974)
Source: www.nlwkn.de

[ha]	The Netherlands	Denmark	Germany			Total
			Lower Saxony	Hamburg	Schleswig-Holstein	
Island	3,420	2,890	2,640	185	720	**9,855**
Main Land	2,190	4,590	5,460		7,500	**19,740**
Summer Polder	1,980		1,795			**3,775**
Halligen					2,300	**2,300**
Total	**7,590**	**7,480**	**9,895**	**185**	**10,520**	**35,670**

Fig. 2.3 Salt marsh areas in the Wadden Sea
Source: Stock (2002) and Ahlhorn and Kunz (2002b).

The present area of salt marshes around the Wadden Sea is in the main the remainder of larger former wetlands. Nowadays, almost all salt marshes are man-made or strongly influenced by human activities like coastal protection works, farming or other kinds of land use. In the middle of the last century salt marsh areas increased. This tendency was lowered up till the end of the 20th century, with a slow increase now detectable due to changes in foreland protection works (Dijkema 1987, Dijkema et al. 2001). Stock et al. (2001) concluded that from 1978 to 1996 the amount of salt marshes increased by about 1,700 ha. However, a large amount of salt marshes was lost by dike-building (approx. 1,500 ha). Today, neither a large decrease nor an increase of the salt marsh area can be detected, but within the groyne fields the tendency towards more erosion increases. Local loss can occur due to poor sediment conditions or erosion in the adjacent tidal flats (Stock et al. 2005). At certain places in Lower Saxony, the area of salt marshes has increased by about 2,747 ha and in other places it has decreased in the same period to 233 ha, so in total an increase of approx. 2,500 ha over the last 30 years can be determined. The recent investigation (in 2003) of the development of the salt marsh area in Lower Saxony gives a detailed overview of the changes, e.g. a detailed description of how the salt marshes have been restored or were lost due to dike-building or agricultural use. The increase of the salt marsh area in Lower Saxony is mainly in the sheltered bays of Jade Bay and Ley Bay (Bunje and Ringot 2003). In the Danish part no clear tendency could be established due to the lack of reliable data series. The comparison of the last available data and the latest investigation shows an increase of Danish salt marshes to approx. 8,710 ha. An increase can be found in some places on the mainland and in the proximity of the island of Fanø (Bakker et al. 2005).

A general tendency is the decrease of the pioneer zone and the increase of the older salt marsh parts which lie above the local mean high water level. This could be the consequence of higher energy input into the Wadden Sea and of the loss of space for natural salt marsh development e.g. due to land reclamation works in the last centuries (Dijkema 1987, Doody et al. 1993, Stock 2002). This effect is called *coastal squeeze*. Recent investigations in the Schleswig-Holstein Wadden Sea confirm these tendencies. The report states that these tendencies may accelerate under climate change: a progressive narrowing of the Wadden Sea, i.e. coastal squeeze will take place under an accelerated sea level rise, an increase of storminess will lead to higher hydrodynamical forces on the sand and mud plates. Consequently, this will lead to a significant loss of specific habitat such as mussel beds and eelgrass (Dolch 2008). The development and the behaviour of the Wadden Sea is essential for both nature conservation and coastal protection. Nature conservation might well get in conflict with the aims of process preservation and habitat conservation; the (natural) process is coastal squeeze; the consequence may be loss of habitat. Coastal protection needs the functions of the Wadden Sea and the salt marsh (foreland). The dissipation of wave energy from the North Sea increases if the sand and mud plates grow, retardation or even decrease will reduce this feature.

Coastal Squeeze

If the sea level rises, as it has since the last ice age, intertidal areas will naturally migrate landwards, maintaining the same position relative to the high and low tide marks in which the plants and animals thrive. If there is a fixed barrier, such as a dike or sea wall, this landward migration is interrupted. This means that the plants in existing areas of salt marsh will die, but no areas of replacement habitat become available further inland because of the barrier. The salt marsh is squeezed out between the sea and the barrier; erosion will appear and the salt marsh eventually may disappear – see for example Mai and Bartholomä (2000), Doody (2004).

van Duin et al. (1999) distinguished three morphological types of salt marshes:

Island Salt Marsh: Three sub-types of almost natural salt marshes can be distinguished on the islands. *Barrier-connected salt marshes* developed at the lee side of sand dune systems of barrier islands. A thin cover of clay-containing layers, starting from a former sandy beach plain, allows the establishment of salt marsh vegetation. The morphology shows an intricate pattern of creeks, levees and basins. Various transitions between salt marsh, beach plain, dune slacks, and dry dune occur and may show a relatively high species diversity. Seawards, they resemble foreland salt marshes. *Green beaches* develop on high and open beach plains. *Foreland salt marshes* develop in front of some island dikes. They are more clayish and richer in organic matter and the clay-containing layer is of a greater thickness than in barrier-connected salt marshes.

Mainland Salt Marsh: Along the mainland coast, two salt marsh types, mostly man-made, can be found. The first type are salt marshes situated in front of the mainland coastal plain, normally bordered by dikes at the landward side. The development has been stimulated by regulation of two key processes: enhancement of the drainage and reduction of wave/current energy. In Denmark and Germany, there are salt marshes along the mainland coastline which are not man-made or influenced by coastal protection measures [...]. The estuarine type resembles the foreland type, but the vegetation and invertebrate fauna show a brackish gradient, perpendicular to the normal zoning.

Halligen Salt Marsh: Halligen are splendid salt marsh islands on dwelling mounds. They have been naturally accreted on surviving parts of marshes flooded in the past, and are highly exposed to wave energy.

In terms of vegetation salt marshes itself are divided into six types. Here the TMAP (Trilateral Monitoring and Assessment Programme) classification is used – see Bakker et al. (2005) and Fig. 2.4:

Pioneer Salt Marsh: This zone lies approx. 40 cm below mean high water level with *Spartina anglica* and *Salicorna* spp. (Samphire) as main vegetation types.

Low Salt Marsh: This zone is inundated during mean spring tide, approx. 100–400 floods per year. The main vegetation is *Puccinella maritima* and *Aster tripolium* (beach aster).

Middle/High Salt Marsh: This zone is inundated less than 100 times per year. The vegetation is dominated by *Festuca rubra* (red fescue) and *Juncus gerardi* (salt rush).

Green Beach, Sandy Pioneer Zone: Mainly found on the barrier islands with vegetation like *Elytrigia juncea*.

Brackish Marsh: Salt marsh zone found in the estuaries influenced by salt and fresh water, with *Phragmites australis* (reed).

Fresh (anthropogenic) Grassland: Former salt marshes truncated by salt water influence due to building of embankments, with *Lolium perenne* (perennial ryegrass).

For the case study Nessmersiel these salt marsh types have been aggregated to three Design Elements (see Sect. 5.2.4): pioneer zone, salt marsh and marsh.

Salt marshes exhibit about 40 typical plant species that are 90% dependent on the special situation of salty ecosystems; i.e. on good nutrition support and also good sun conditions. About 1,650 terrestrial animal species and approx. 350 marine animal species live in the salt marshes, half of which are strongly connected to salt marshes (Heydemann and Müller-Karch 1980). Decreasing influence of salinity and flooding causes an increase in plant species (Heydemann and Müller-Karch 1980). Salt marshes are a roosting, feeding and moulting area for many birds (see Box on p. 19). Some of these birds are listed in the Red List of Lower Saxony and Bremen, e.g. the lapwing and the redshank are classified as endangered or highly endangered (Südbeck and Wendt 2002). Besides that, salt marshes fulfil other important functions such as filtering North Sea water and they have an asthetic value (Heydemann 1987). To evaluate the quality of salt marshes Dierßen (1987) recommended five criteria: representativeness of areas, the sparseness of existing or resettled species of plants and animals, the diversity and variety of existing or new spatial structures and the status-quo of each area. The importance of the salt marshes for nature conservation and coastal protection in respect of the functional value have been thoroughly investigated by Meyerdirks (2008) and Wittig (2008). A detailed explanation of the functional value is given in Sect. 4.3 on p. 110: the functional

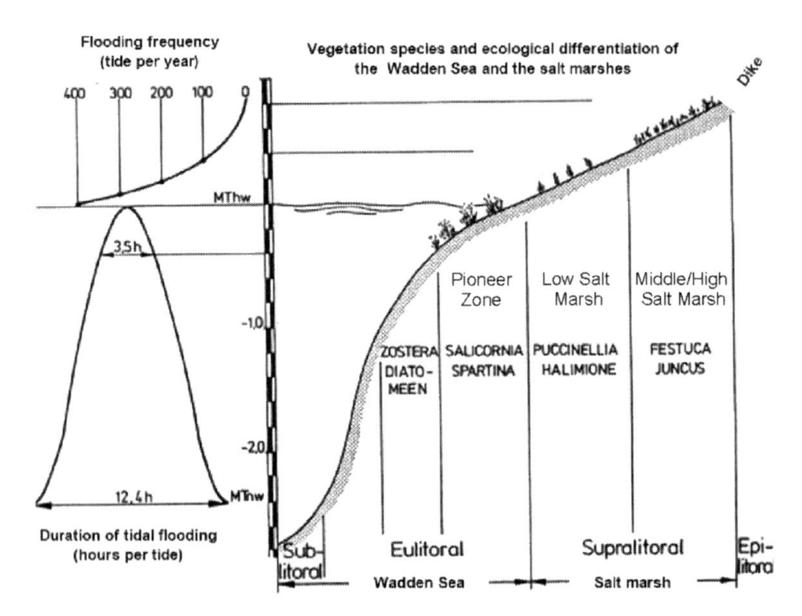

Fig. 2.4 Interaction of tide level and salt marsh vegetation
Source: Bretschneider et al. (1993).

value describes the services which a natural unit (here the salt marsh) provides for different types of land use.

The quality of the salt marsh and the effects of coastal protection schemes on salt marshes have been investigated within several projects. Within these projects different items have been investigated e.g. protection of biotopes, protection of species and the potential for development of areas with regard to changes in reclamation efforts and land use (Michaelis 1968, Arens and Götting 1997). The importance and the quality of salt marshes has been comprehensively investigated, e.g. as conservation of evidence following coastal protection projects see e.g. Heydemann (1987), Blindow (1991), Arens (2000), Götting et al. (2002) and within the German research programme "Climate Change and the Coast" see e.g. Kinder et al. (1993), Cordes et al. (1997) or Vagts et al. (2000).

The interest of nature conservation in maintenance and development of salt marshes are described in the "mission statement" for the Wadden Sea National Park in Schleswig-Holstein: A salt marsh not used by human beings with natural channels and ditches, characteristic and geomorphological structures and a characteristic distribution of plant and animal species regarding the natural dynamics – see Stock et al. (1994). For Lower Saxony the salt marshes are very important in the Wadden Sea area and are highly protected, but no "mission statement" for the salt marshes has been developed.

The Wadden Sea Region – Important Bird Area

The Wadden Sea attracts about 50 bird species with more than 10 million individuals which breed, rest and some of them stay over winter-time. The Wadden Sea is attractive for birds because of the high production rates of biomass and a good availability of nutrients. The common breeding birds are black-headed gull, herring gull, arctic and common tern and avocet (Exo 1994).

Breeding Birds

> ... The Wadden Sea is a hot-spot within the European breeding range and which represent Species of European concern. Furthermore, 14 species are included in Annex I of the EC Birds Directive (EC 1979) and several breeding birds are listed in national Red Lists for Denmark, Schleswig-Holstein, Niedersachsen or The Netherlands. The distribution of breeding birds within the Wadden Sea is mainly determined by geographical range, feeding opportunities, available nesting habitat, predation pressure and level of human disturbance. High densities of breeding birds are especially found in salt marshes, the dunes on the islands and the higher outer sands (Koffijberg et al. 2005, p. 275).

Several factors influence the occurrence of breeding birds in the Wadden Sea: climate change, pollution, recreation and tourism, fisheries and agricultural use of salt marshes and adjacent resting and breeding areas. The effects of climate change and the accompanying consequence of an accelerated sea level rise might have negative influences on breeding birds, especially on species which breed in the surf zone of beaches and salt marshes. The influences of milder winters might have positive as well as negative effects. Recreation and tourism will have a negative influence, since it takes place in the area of breeding birds, especially at the surf zone. Changes in management have dampened the influence in some places. Fisheries have caused an ambivalent effect on breeding birds. On the one hand, the population of some species has grown because of the increasing fishery discard, and on the other hand the negative effects of harvesting mussel beds. Ambivalent influence can be determined for agricultural use of salt marshes. The change of salt marsh management has led to a more natural development of salt marshes accompanied by higher vegetation. Consequently, the composition of the species has changed to birds which are more adapted to higher vegetation like redshank and meadow pipit (Koffijberg et al. 2005).

Migratory Birds

The Wadden Sea area plays an outstanding role for migratory birds enroute to their breeding range or on the way back to their wintering areas. Within the Quality Status Report 2004 (Essink et al. 2005) 34 species were included in an evaluation of the state of migratory birds in the Wadden Sea: 44% showed a significant decrease in the 1990s, another 21% showed a decrease which is statiscally insignificant. These are not common tendencies, in adjacent areas like the UK and France these tendencies were not observed, thus the reason for decline of species might be found in the Wadden Sea. A few species have increasing trends, like barnacle goose and eurasian spoonbill. Others have fluctuating trends, because of low abundance (Blew et al. 2005). Most migratory birds do not only roost, but also moult in the Wadden Sea. Within this moulting time the birds are highly vulnerable, because many of them can not fly. "Case studies in several parts of the Wadden Sea have pointed out that recreational activities are among the most frequently observed sources of anthropogenic disturbance. This is confirmed by the recent inventory by Koffijberg et al. (2003), which points out that 29 to 42% of all roosting sites are subject to an estimated moderate to heavy recreational pressure." (Blew et al. 2005, p. 292). The highest abundance of moulting shelducks can be found in the southern part of the Schleswig-Holstein Wadden Sea with about 200,000 individuals. The moulting areas of the common eider is not concentrated like the moulting area of the shelduck, but account for about 170,000–230,000 individuals. Remarkably, in the East-Frisian region between the barrier island of Juist and Wangerooge there are no roosting sites for these species. Blew et al. (2005) concluded that within this area the recreational pressure and activities are too high to provide undisturbed areas for roosting and moulting birds. Other disturbances for roosting and moutling birds are commercial fishery, boat and air traffic and oil spills. Additionally, in the future, near-shore wind farms within the 12-nautical-mile-zone might prove potential elements for disturbance.

2.2 Sea Level Curves and the Flood-prone Areas

2.2.1 Sea Level Curves

In this section a brief description will be given of the recent findings related to the extent and the reasons for sea level rise. The morphological structure of the southern North Sea coast is mainly dependent on these processes. The iso-static movements in the southern North Sea are mainly caused by the retreatment of the ice mass since the last ice-age and reflect the balancing of the earth's crust. The eu-static sea level rise is also caused by the retreatment of the ice mass, since the melting ice led to an increasing sea level. Many investigations have been carried out world wide and documented in a wide range of publications. A current overview is given by the Intergovernmental Panel on Climate Change (IPCC) – e.g. IPCC (2007). A frequently cited curve which visualises the quaternary history of sea level changes in the North Sea has been published by Jelgersma (1979).

A new sea level curve for the southern North Sea by Behre (2007) is shown on Fig. 2.5. The sea level curve is mainly deduced from archaeological data. The sea level rise amounted to 1.25 m per century from 7,000 to 5,000 BC, and was reduced to 0.14 m per century until 1,000 BC. Between 1,000 BC until today the curve shows many oscillations. These oscillations are explained by information and data on settlement activities in the low-lying areas of the German Bight. In mean, the curve shows over the period of the last 3,000 years a sea level rise of about 0.11 m per century.

Bungenstock (2006) investigated the sea level curve of the barrier island of Langeoog (Lower Saxony) using the sequence-stratigraphy method. This curve is inserted into Fig. 2.5. The sea level curves show more or less the same rates, although two different approaches were used, the natural-scientific approach by Bungenstock (2006) and the archaeological approach by Behre (2007).

The sea level curves presented give an impression of the sea level rise for the last 10,000 years. However, the determination of the sea level rise and thus these curves are imprecise due to different reasons. Behre (2003) refers to these aspects and discussed these reasons, e.g. determination of the age using archaeological data, e.g. elevation of settlements, or determination of the height using peat. The age of archaeological data is subject to estimations. The height of the mean high water level has to be determined carefully by using the peat horizon, because some turfs can develop between 0.5 m below mean high water level (MHW) and MHW and others between MHW and 0.8 m above MHW. So, there is a potential divergence of approx. 1.3 m. Using the elevation of settlements on ground level is also imprecise because these settlements represent more or less storm surge level rather than MHW, and have been set as 1 m above MHW (Behre 2003). The same problems of interpretation e.g. of the heights of peat apply to Bungenstock's approach. Behre (2003) concludes at present a tectonic decrease of less than 0.1 cm per century and that a further iso-static movement is unlikely. Bungenstock (2006) concludes that the coast of the southern North Sea is – in terms of geological time scales – a sedimentation

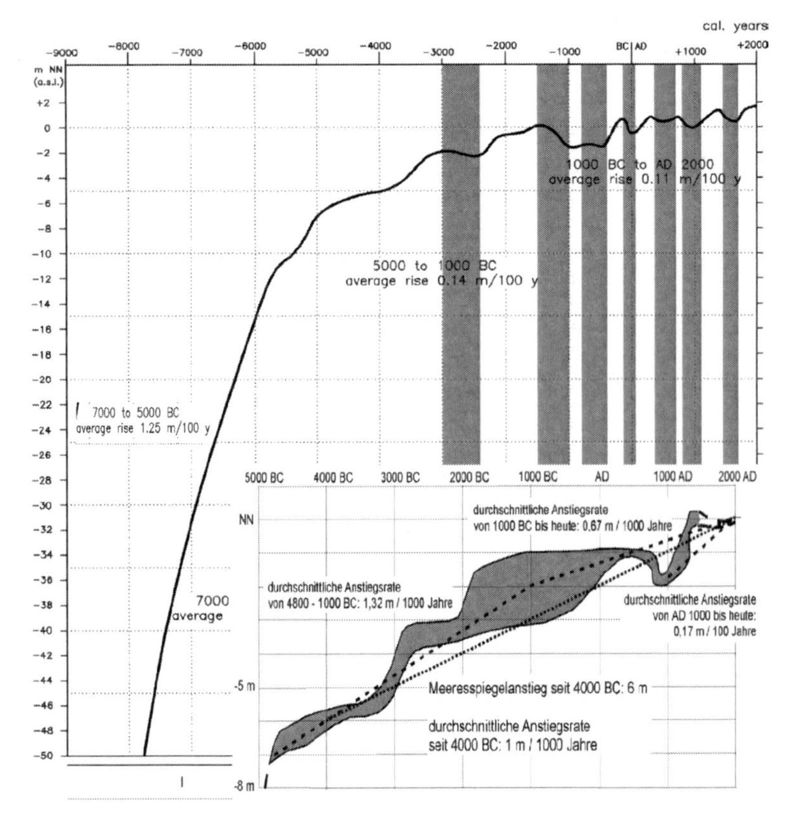

Fig. 2.5 Sea level curve for the southern North Sea
Source: Bungenstock (2006) and Behre (2007).

coast. However, short periods (about a generation of human live) of sea level rise cannot be ruled out.

2.2.2 *Todays Flood-prone Areas*

In Germany the States are responsible for coastal protection. An overview of the flood-prone area in the German Bight is given in Fig. 2.6. In *Schleswig-Holstein* the existing Master Plan for Coastal Protection Management defines the area between the −15 m NN on the sea-side and +5 m NN on the land-side as the coastal protection planning area. The flood-prone area is about 3,400 km^2 with approx. 250,000 inhabitants. The sea-side area covers approx. 3,000 km^2 including the islands, Halligen and the off-shore island Helgoland. The Master Plan of Schleswig-Holstein covers both the land and the sea side. The flood-prone area is determined by the area which will be inundated by the highest storm surge level without protecting elements. The sea side border was chosen because of the influence of waves and currents (MLR 2001).

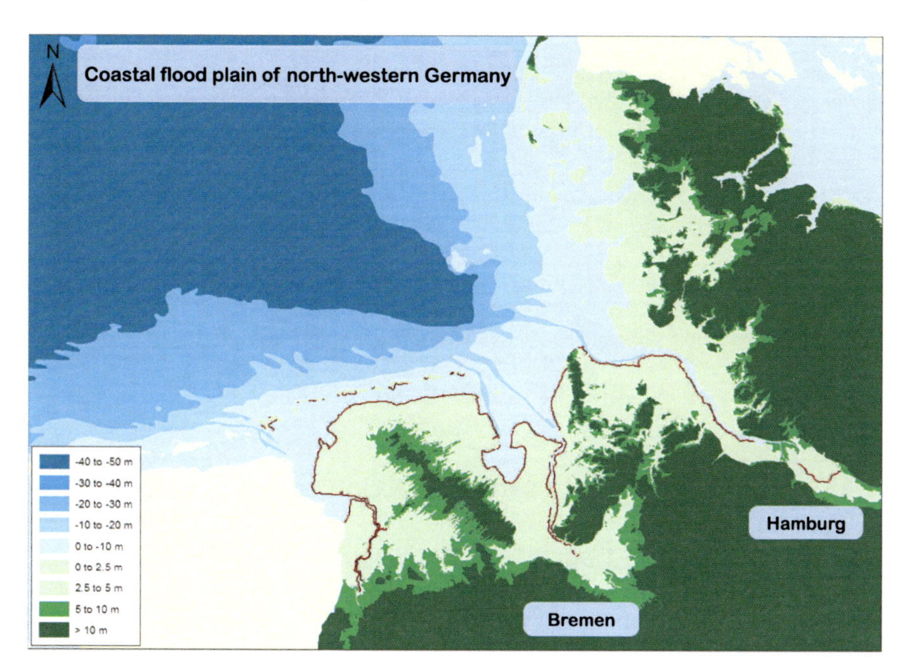

Fig. 2.6 The coastal flood plain of north-western Germany
Source: Ebenhöh et al. (1996).

Regarding the State Law on Dikes for *Lower Saxony* the protected area (=flood plain) is the area which will be inundated by the highest storm surge level (§6, NDG 1963). This area is the area of the dike boards along the Lower Saxonian coast and covers approx. 6,500 km^2 (MELF 1973, BR W-E 1997) with approx. 1.2 million inhabitants. On the sea side, the area is limited by the MHW, because it is the natural border between the fore land and the Wadden Sea which is flooded by normal tide twice a day (Lüders and Leis 1964). In Lower Saxony the area for coastal protection ranges from the MHW to the NN +8 m contour-line in the hinterland. The barrier islands are located in this area, but they are outside the scope of this dissertation.

The flood-prone area of the state of *Bremen* is also determined by the highest storm surge level. For some parts of the city of Bremen the river discharges (run off) have to be taken into account (e.g. River Weser, Ochtum and Lesum). The flood-prone area is approx. 340 km^2 with approx. 410,000 inhabitants (NLWKN 2007b). In *Hamburg* one third of the city is lying below MHW, approx. 250 km^2, with approx. 180,000 residents and 140,000 employees (Otto 2004, LSBG 2007).

The figures summarised in Fig. 2.7 are mainly taken from the technical reports of the EUROSION project – see EUROSION (2004a,b). Within the EUROSION project a “Radius of Influence of Coastal Erosion := RICE” was defined: the area within 500m from the coastline that can be extended to areas lying under +5 m MSL. In England the flood prone area is determined as an area possibly affected by a 1 in 5 year storm event and a 1 in 200 year storm event, respectively, which might exceed the +5 m contour-line (EUROSION 2004b).

Items \ Country	Germany				East of England[5]	Netherlands[5]	Denmark[5] (North & Baltic Sea)	Total [approx.]
	Lower Saxony[1]	Schleswig-Holstein[2]	Bremen[3]	Hamburg[4]				
Flood Prone Area [km²]	6,500	3,400	340	270	3,714	13,900	13,300	**40,373**
Inhabitants [approx.]	1.2 Million	252,000	410,000	180,000	855,000	7.1 Million	2.1 Million	**12.1 Million**
Border of the protected area	up to +8m contour-line	+5m contour-line	+5m contour-line	+5m contour-line	Line determined by 1:200 storm event	+5m contour-line	+5m contour-line	X

Fig. 2.7 Figures about the coastal flood plain of the southern North Sea region
Sources: 1: MELF (1973), BR W-E (1997), NLWKN (2007b) 2: MLR (2001), 3: NLWKN (2007b), 4: Otto (2004), LSBG (2007) and 5: EUROSION (2004b).

2.3 Service for Society: Coastal Protection

2.3.1 Retrospective – Coastal Protection until Yesterday

The purpose of this section is not to repeat the description of the historical development of the organisation and engineering techniques of coastal protection. A comprehensive description of the dike-building techniques can be found in e.g. Kramer (1992). But, to understand the development of coastal protection in north-western Germany, especially in Lower Saxony it is necessary to describe roughly the historical changes until the severe storm surges in the middle of the last century. Comprehensive descriptions of the historical organisation and the circumstances of coastal protection along the coasts of the southern North Sea can be found e.g. in van de Ven (1993). The historical development in Germany, especially in the States of Lower Saxony and Schleswig-Holstein has been treated in several reports and articles. For the purpose of this report it is sufficient to refer to some old literature like Brahms (1754), Auhagen (1896), Tenge (1898), von Gierke (1901/1917), Tenge (1912), Wöbcken (1924, 1932) and Breuel (1954), which most of the following and recent publications are based on. Nevertheless, the research and investigations into early settlements in the coastal zone has provided a deeper insight into the challenges for living and working in low-lying areas – e.g. Krämer (1984), Hofmeister (1984), Brandt (1984), Prange (1986) and Behre (1987).

People settled down in the low-lying areas of the southern North Sea region and tried to provide for their livelihood. The first settlers in the low-lying areas of the north-western part of Germany constructed their settlements according to the sea level (i.e. intuitive storm surge level). This defensive strategy (or adaptation), i.e. moving inland with rising water and moving towards the sea with falling water, was reconstructed by excavations in the lower marsh (Brandt 1992). The settlements in the early middle ages were limited to the higher grounds along channels and rivers. Afterwards, in times with low changes in sea level, the re-settlement of the area led to the protection against storm surges of the sea by building dwelling mounds – e.g. Brandt (1984, 1992). In the last period of the 12th century and with the beginning of the 13th century the first embankments were built, to protect the land around settlements (Brandt 1992). These embankments displayed a very poor construction in comparison to today's, and can perhaps be compared to current summer dikes, which are only capable of protecting the farmland against lower high-tides occurring during summer time and autumn (Peters 1992).

To conclude: The first embankments were built in the earliest mid-ages around the southern North Sea (e.g. van der Linden 1981, Blok 1984 and van de Ven 1993 for The Netherlands and Ravensdale 1981 for England) to reclaim land for settlement. Thus, the fight against the sea was the main task for these low-lands, expressed by the Frisian settler: "Frisian people should protect their land with three weapons: spade, pushcart and pitch-fork" (p. 6 after Wiarda 1805 in Meyer 1926). The contract in the year 1106 between settlers from The Netherlands and the archbishop of Bremen seems to be the first hint of dike-building to enable settlement at the river marshes of the Weser (von Gierke 1901/1917) and thus to reclaim arable land.

For the purpose of this dissertation, especially for the issue of participation, it is of special interest how the alliances or unions to protect and to maintain the reclaimed land were established and how they were organised and managed. These characteristics provide a deeper insight in the traditional identity and mission of dike boards today. The land reclamation works in the fenland's of The Netherlands, the area between Utrecht, Rotterdam and Amsterdam, had in earlier times led to the generation of local organisations to commonly maintain the already installed constructions (i.e. dams, sluices and dikes). These organisations consisted mainly of the aldermen of the settlements. These organisations developed over time and grew with the reclamation works and the interests of maintenance and drainage of a much bigger area. The water boards in The Netherlands were established and officially commissioned by a count or a bishop in the 13th and 14th century. Afterwards, these water boards also became important for the creation of polders, which were mainly designed to control the water level of the arable land inside a dike-ring; the oldest polder can be found in Zeeland. Starting in 1840, the management board of the water boards underwent several changes until the end of the 19th century. At that time the provinces became more powerful and published new laws and regulations. These changes led to more democratic structures, based on assemblies of representatives elected by the landowners (van der Linden 1981). The number of water boards in The Netherlands was reduced from over 3,500 to 2,500 by 1950 and after the devastating storm surge in 1953 to about 120. Nowadays, there are 26 big water boards in The Netherlands with comprehensive duties such as the maintenance of dikes and other embankments (in The Netherlands about 3,500 km), water management, water quality and to some extend for waste water treatment (UvW 1992).

The establishment and the historical development of the dike boards in Germany are outlined by Kramer and Rohde (1992), and e.g. for the Oldenburg-area by Meyer (1926). The dike boards in north-western part of Germany were established in different ways, but the result was always the same: the dike boards were organisations set up for a special reason: protection of settlements and arable land. The early dike boards were associations like municipalities with obligatory membership within a specific area. During the 16th and the 18th century the organisation of the dike boards was reformed, the associations were compelled to transfer more and more responsibilities to state agencies. Hand in hand with these changes manual dike construction work was replaced by machine-assisted techniques. Also a system of fees was introduced. After the storm surge of 1825 these developments were rebirth of the adoption of the co-management. The decisive step in the re-reform process was the dike order of 1855, which was based on four principles (Meyer 1926, p. 101):

1. Establishment of a common dike law for the entire state.
2. Reduction the number of dike boards: from 15 to 4.
3. Annulment of exoneration of dike obligations.
4. Annulment of the deposit on the past of the inhabitants protected by main dikes, and adoption of the "Kommuniondeichung" (which distributed dike maintenance duties among all people living in the flood-prone area of a dike board).

The latter principle was a reaction to the bad experiences made in earlier times, when the responsibility for a line lay solely with the people living directly behind a certain stretch of dike with the maintenance and reconstruction was based on a system of deposits. The old approach led to the desolate economical state of the dike boards. Finally, the process was stopped in the last century with a democratic organisation of the dike boards, run by a committee, where members were elected (Meyer 1926). To conclude, the result of joint work for coastal protection was the generation of marsh-land, which led to economical growth.

The difference between the dike boards in Germany and the water boards in The Netherlands today lie in the distribution of the responsibilities and their place within the government/authority hierarchies: the "water boards" are also responsible for water management of the area, whereas the "dike boards" are solely responsible for the dikes. The water boards in The Netherlands have an equal status to that of the municipalities, in Germany the dike boards are public corporations with a special duty, which in Lower Saxony is laid down in the State Law on Dikes (NDG 1963).

2.3.2 *The Consequences of the Storm Surges in 1953 and in 1962*

The storm surges in 1953 in The Netherlands/England and a few years later in 1962 in Germany led to both organisational and technical changes, i.e. in the fields of engineering and the strategic orientation of coastal protection. The following section summarises the circumstances, experiences and the consequences of these devastating storm surges. Again, many publications are available on the consequences of these storm surges and in the light of the recent storm surges at the German coast, these aspects have been reviewed in several documentations. A comprehensive overview is given e.g. in Kramer (1989), Kunz (2004a) or DHV (2007).

2.3.2.1 Northern Germany – Lower Saxony in 1962

In Germany, especially in Lower Saxony, the strategy for the mainland can be summarised as "hold the line". As a reaction to the storm surge of 1953 in The Netherlands an Engineering Commission was established to investigate and to improve the existing strategy of coastal protection – see e.g. Lorenzen (1955), Tomczak (1955). The recommendations led to higher crests on some coastal stretches which, later withstood the water heights and the wave run-up of the 1962s storm surge (Ingenieur-Kommission Niedersachsen 1962). The storm of 1962 led to extreme water levels in the area of the rivers Elbe and Weser, especially in Hamburg and Bremen, and in the river Ems area. The "Lower Saxonian program for coastal protection 1955–1964", installed after the Dutch disaster, was not completely implemented by February 1962. The consequences were loss of life of people and animals and devastating damages, mainly in Hamburg. The water levels at the East Frisian coast were less than expected. The main dike line extended approx. 870 km along the Lower Saxonian coast; about two thirds of which were not damaged, but the consequence of 61 dike breaches was an inundated area of approx. 37,000 ha (Ingenieur-Kommission Niedersachsen 1962). The Engineering Commission acknowledged that older dike lines, which had been built with a different mission (i.e. to protect the polder against inland waters from the peaty area), had protected the hinterland against devastating inundations. Thus, the recommendation was given to maintain and improve the older dike lines, and where possible to build new second dike lines. Also, it was recommended that the second dike line should be installed within a certain distance of the main dike to enable better emergency management in the case of a storm surge and a dike failure. The experiences in 1962 have shown, that the response time in some cases was too short to rescue the inhabitants of the polder areas (Ingenieur-Kommission Niedersachsen 1962).

The experiences of the 1962-disaster led to the installation of the Advisory Committee for Coastal Protection for the North and the Baltic Sea (Lorenzen 1966). The comprehensive recommendations of this Committee were published in Engineering Committee for North and Baltic Sea (1962). These recommendations cover the design of dikes, the design of the outer and inner slope, the quality and conditions of the soil and the subsoil of dikes, and the maintenance and contingency planning. The results of the research programmes and the recommendations of the committee led to the first-ever State Law on Dikes in Lower Saxony – see NDG (1963), a detailed description of which will be given in Sect. 4.1.1.

The coastal protection authority is obliged to prepare a special plan describing and elaborating the strategy and concepts of coastal protection (Master Plan for Coastal Protection – German: Generalplan Küstenschutz). The first Master Plan, published in 1973, included a comprehensive description of the consequences of the storm surge in 1962, objectives of coastal and island protection in Lower Saxony and the design of main dikes and their heights, and also the work programme for the coming years – see MELF (1973). The intention of the Master Plan is to fix mandatory directives and to provide information for both the general public and the people and institutions affected.

2.3.2.2 The Netherlands since 1953

The consequences of the storm surge of 1953 (or "de ramp") were the loss of life of about 1,800 people and many animals, because of several dike failures and the resulting devastating damages (e.g. Slager 1992, TAW 1998, Seijffert 2001). As a reaction the strategy of coastal protection was totally re-worked for The Netherlands (TAW 1998). A master plan to increase the safety of the low-lying areas was established in the 1960s. The basis of these coastal protection works is the Flood Defence Act (FDA) established in 1953. Up until the 1950s the risk of flooding was estimated on the basis of intuition and experiences, as (complex) simulations or calculations to determine the risk of flooding were not possible (TAW 1998). The first step of the advisory board of the Delta Committee was to investigate whether the water level of the 1953 flood (NAP +385 m, Hoek van Holland) could be exceeded and the second step was to investigate the costs of increasing the safety level in comparison with the expected economic benefits (TAW 1998, 2000). The result was the classification of the Dutch coastline and the rivers into four safety levels, i.e. ranging from 1 to 10,000 for highly populated areas with high economic values from 1 to 4,000 at the Wester Schelde delta and the north-east of The Netherlands, and 1 to 1,250 at the rivers (e.g. de Ronde et al. 1995, TAW 2000). The proposed coastal works, called "Delta Works", were largely completed in the late 1980s. This approach was based on the suggestions to strengthen the embankments as a result of the experiences of the storm surge in 1953 in the light of the state-of-the-art in engineering, i.e. the prime goal was to retain the water from the hinterland. Hence the embankments were constructed to withstand one major flood every 10,000 years.

This probability-based approach of exceedance has been applied to the entire coast of The Netherlands. Over the past 30 years many dikes, sluices and storm surge barriers have been improved – and to some extent newly installed – to increase safety against flooding. However, the design of the embankments was subject to the lack of experience and knowledge of the various failure mechanisms, which could lead to a dike failure (TAW 2000). The Technical Advisory Committee for Flood Defence published comprehensive guidelines for the maintenance and the construction of embankments and the consideration of different values and functions in the coastal zone and along the rivers – see TAW (1998). Shortly after the completion of the Delta Works a new policy of dynamic maintenance was established, accompanied with the base-line concept for the sandy coast in The Netherlands (RWS 1990, Hillen and de Haan 1993). The new strategy was a reaction to the enduring coastal erosion along the coasts of The Netherlands and the anticipation of possible consequences of an accelerated sea level rise. In general terms: if the base-line concept was applied; erosion can be allowed to take place to a limited extent, but will be prevented on highly vulnerable coastal stretches. The main technique deployed to implement this strategy was the application of sand nourishment and this has led to reasonable results. The concept of the coastal defence strip was introduced from the −20 m depth-line on the sea-side to the polder on the land-side (MVenW 1990).

The Netherlands applied a scenario-driven approach for an accelerated sea level rise, which is based on the IPCC scenarios and ranges from 20 cm/100 years (minimum scenario) to 85 cm/100 years (maximum scenario) with 10% more winds (MVenW 2002). The safety level of the embankments will be re-assessed every five years. The Technical Advisory Committee recommended that the failure mechanisms and the probability of embankment-failure be investigated and included in the planning measures (TAW 1998).

2.4 Protection Against Flooding – Today's Concept

2.4.1 Lower Saxony

Coastal Protection in Germany is based on the Constitution of the Federal Republic of Germany (German: Grundgesetz). In article 74 section 1 coastal protection is an integral part of the concurrent legislation between the Federation and the States (Deutscher Bundestag 2007a). In practise, the responsibility lies with the States (principle of subsidiarity). All coastal States have a special legislation for water management and all of them have incorporated coastal protection into this framework with the exception of Lower Saxony with its State Law on Dikes (NDG). The Federation contributes financial support to this task, established in article 91a of the Constitution (Deutscher Bundestag 2007a). The contribution is established as "Federal Objection for the Improvement of the Agrarian Structure and Coastal Protection". The present version refers to the main principle for the enhancement of the agrarian structure and coastal protection for the period 2007–2010 (Deutscher Bundestag 2007b). The fields of enhancement encompass e.g. improvement of rural structures, improvement of production and distribution structures, forestry and coastal protection. The aim for coastal protection is as follows: "Defence against natural hazards and enhancement of safety in coastal zones, on the islands and on river basins in tidal areas against inundation and loss of land by storm surges and sea attack" (Deutscher Bundestag 2007b, p. 62). Other provisions are included, such as strategy development and investigations in combination with measures, to reconstruct, strengthen and heighten coastal protection structures, fore land work within a range of 400 m and sand nourishment. Limited support is given for coastal protection measures, which affect areas of ecological value. These are only eligible if e.g. the required safety can not be achieved by another justifiable measure. Measures which are not covered by the Federal Objection are e.g. the maintenance of coastal protection structures and the construction of pumping stations. The federal contribution is limited to 70% and the State contributes 30% (Deutscher Bundestag 2007b).

The recommendations of the Engineering Committee for Coastal Protection of the North and Baltic Sea are still valid (Engineering Committee for North and Baltic Sea 1962) and have been amended in the light of recent findings in 1993 (EAK 1993) and 2002 (EAK 2002). The latest Master Plan for Coastal Protection in Lower Saxony in combination with Bremen (Mainland) was published in 2007 (NLWKN 2007b, a detailed description of which is given in Sect. 4.1.1).

The concept and the strategy of coastal protection has not been changed since the Master Plans of 1973 and 1997. Hence, the current Master Plan is an extension of former Master Plans, describing mainly the financial stipulations and the agreed programmes for the next years. The budget claimed by Lower Saxony is approx. 500 Mio. € (NLWKN 2007b). The necessary projects will be described in more detail in Sect. 3.4, because these projects relate directly to activities in Lower Saxony.

The methodology for calculating and determining the height of the dike is recommended by the Advisory Committee for the North and Baltic Sea and combines two approaches: (a) Single Value Proceeding and (b) Composition Proceeding (Fig. 2.8).

Although Lower Saxony features second or older dike lines, these are not adequately considered and to some extent they have not been integrated into the concept of the protection by the dike boards (Kunz 2004a). Examples of different second dike line conditions are demonstrated by Fig. 2.9: a second dike line of the Norden Dike Board is shown in Fig. 2.9a; the dike board responsible for the coastal stretch around the western Jade Bay (III. Dike Board Oldenburg) has improved and maintained the existing second dike lines over the last years – see Fig. 2.9b.

The amendments of the Master Plans conducted and published after the storm surge of 1962 follow the recommendations of the Advisory Committee for Coastal Protection (Kunz 2004b). The aims of safety for coastal protection at the North and Baltic Sea are summarised as follows:

> The basic findings of the Advisory Committee for Coastal Protection – see EAK (2002) – are as follows:
>
> - The absolutely highest storm surge can not be determined.
> - Certain specific astronomical, meteorological and oceanographical circumstances coincide may leading to higher storm surges as anticipated today.
> - It is questionable whether dikes should be built which will hold back flooding water under all perceivable conditions.

This led the Advisory Committee to the following recommendations for a main dike (Kunz 2004b, pp. 256–257):

> - The height of the dike is crucial for the safety of the protected marshes.
> - The crest of the dike should be as high as the design water level, which is based on a defined level of safety taking into account a certain probability of exceedance and the height of wave run-up, which should prevent frequent and powerful overtopping.
> - The decision not to determine a "highest storm surge", has to be considered by designing dikes with a certain cross-section able to withstand strong and long-lasting overtopping.

For the design of main dikes the Committee recommends:

- For every stretch of the coast the "design water level" has to be determined – verified by scientific research and clearly defined – which either will never or at least within a manageable risk level – be exceeded within a certain time period. The currently applied procedures (in Schleswig-Holstein with regard to the frequence

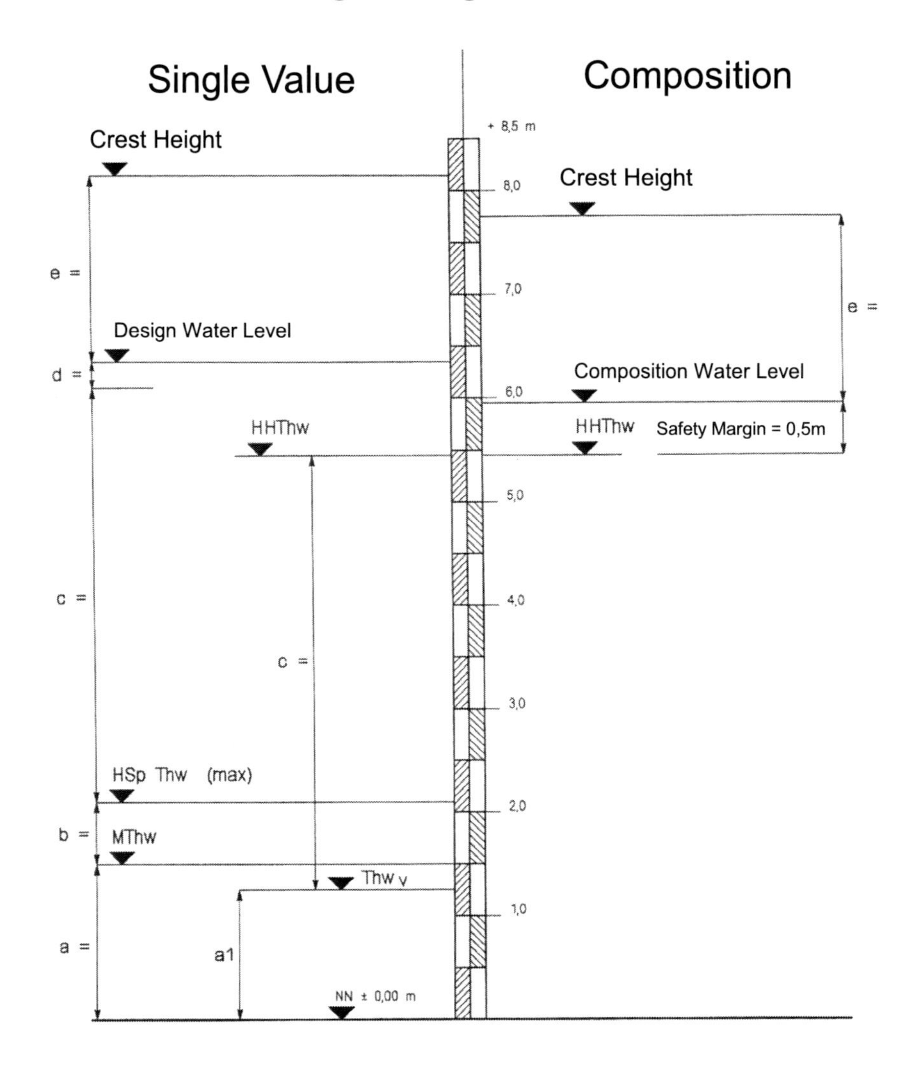

a = Mean High Tide Water (MThw, 10 year mean),
a1 = Forecast Tidal High Water (Thw) with max. wind surge (Windstau)
b = Difference between High Spring Tide Water (Hsp Thw) and Mean High Tide Water,
c = Difference between highest High Tide Water (HHThw) and Mean High Tide Water,
d = Secular trend of Sea Level Rise,
e = Max. wave run-up

Fig. 2.8 Proceedings to calculate the design water level and the crest height of the main dike, *left*: single value, *right*: composition
Source: BR W-E (1997).

(a) Second dike line dike board Norden

(b) Second dike line III. dike board Oldenburg

Fig. 2.9 Second dike lines in Lower Saxony
Source: Ahlhorn (1997, 2005) @ Frank Ahlhorn.

of a certain water level, and in Lower Saxony to the single value procedure) produce approximately the same values, but are not verified by science and are more or less the result of empirical studies.
- Identification of a "design height for wave run-up", which will only be over topped by a certain wave level or only allow a defined amount of water to flow over the dike crest (overflow).

These guidelines of the Advisory Committee are reflected in the current procedure to determine the height of the dike crest and, thus, in the accepted safety standard, incorporated in the procedure applied in Lower Saxony as shown in Fig. 2.8. Differentiated discussion and criticism has been provided by Kunz (2004b) and can be summarised as follows: Since no absolutely safe coastal protection against flooding is guaranteed, the risks have to be limited and managed. There are many uncertainties which have to be considered to determine the safety level of embankments. As mentioned above, for example, the highest storm surge can not exactly be determined, e.g. because of uncertainties in projecting water levels into the future (see box in Sect. 3.1.3). There is no question, that the existing safety concept has to be enhanced, however, step by step and not immediately by introducing new approaches before they are ready for implementation, e.g. it is necessary to enhance the knowledge of failure mechanisms and the failure probability of dikes and other structures (Kunz 2004b).

Taking these remarks into account, the recent enhancement of the safety margin for the secular sea level rise from 25 up to 50 cm announced at the conference on climate change and coastal protection in Oldenburg (July 2007) by the Minister of the Environment, responsible for coastal protection, is a postponement of the issue at stake and does not solve the described basic problems and challenges. However, it serves as an approach to address the consequence of increased sea level rise in Lower Saxony (and Bremen).

2.4.2 The Netherlands

The Technical Advisory Committee in The Netherlands has stimulated an investigation of the failure mechanisms of coastal protection elements. Consequently, a large research project has been launched called VNK (Veiligheid Nederland in Kaart, Flood Risks and Safety in The Netherlands, Floris). The approach taken in this project is described as: "From probability of exceedance to probability of flooding" (TAW 2000, MVenW 2005a–c). This means, that the safety margins established by the Delta Commission are still valid, but the design water level as criteria for the dike height will be substituted by an acceptable risk.

The definitions are as follows (TAW 2000, p. 8):

Exceedance Frequency: The exceedance frequency of a water level is the probability that the design water level is reached or exceeded. The design water level is used to design a safe dike or hydraulic structure.

Flood Probability: The flood probability is the probability, that an area might be inundated, because the water defence around that area (i.e. a dike ring) fails at one or more locations.

The background of the conceptual shifting in The Netherlands refers to the fact, that about 50 dike rings around polders have been installed to protect people against flooding and at the rivers against inundation (some stretches of these dike rings are on rivers and some at the coast). The safety of the dike ring is provided by different constructions like dikes, sluices and locks. Every element has its own failure mechanism and its specific failure probability, especially, that of human error, as the experiences of 1953 have shown (Slager 1992). The aim of the research project was to determine as exactly as possible the probability of failure of these components. This was done by assuming that, a chain is as strong as its weakest link. Thus, the VNK project identified the "Weak Spots" (see Fig. 3.18). In the first phase of this project pilot studies were conducted, see e.g. Provincie Noord-Holland (2005) or RIKZ (2006). This approach and the paradigm shift had been integrated as early as possible in a much wider approach to coastal development – see MVenW (2002). For that purpose, a detailed research programme called SBW (Sterkte Belasting Waterkeringen, Extent of Hydraulic Impacts to Embankments) has been initiated – see e.g. RWS (2008a–d). This programme encompasses six projects:

- Wave overtopping and the strength of the inner slope
- Dune erosion
- Protection of the dune foundation
- Hydraulic boundary conditions at the Wadden Sea
- (Macro) stability of dikes with regard to increased water pressure under the dike
- Wind- and wave statistics.

The new concept of The Netherlands can be summarised as follows: It is important to consider the requirements of water management adequately in spatial planning. The new water management law will strengthen the relationship between water management and spatial planning. This will be implemented by the national and regional water management plans which are part of the new law for spatial planning (Wet over Ruimtelijke Ordening, Wro), which came into force 1. July 2008. With the implementation of this law the spatial aspects of water management will be mandatory at national and provincial levels (description of the integration of water management and spatial planning on www.helpdeskwater.nl of April 2008).

2.4.3 Outlook

The descriptions and explanations in this chapter show that the people living and working in the flood-prone area can protect themselves against flooding. The experiences of the last decades provide evidence, that devastating consequences of storm surges can be avoided. Protection strategies have been revised continuously, e.g. in The Netherlands over the last 20 years, and further aspects have still to be considered. The present strategies will protect people and their property against flooding, but absolutely safe protection against natural hazards is unachievable. Thus, the concepts and strategies have to and will be continuously revised and adapted to new challenges and problems. One of today's widely accepted challenges is linked to the consequences of climate change. Climate change will impose different threats and changes to parameters which are important for the safety of people in low-lying areas and for coastal protection.

Sea level rise itself is not a new challenge for the people living in flood-prone areas, as the discussion of the sea level curves has shown. But, how fast will the sea level rising be? Over the last centuries, we have increased settlements in coastal zones, for the purposes of trading, recreation etc.; but how will the future risks to settlements develop? And what about the Wadden Sea, the enormous nature reserve in front of the dikes? How will the plates react to sea level rise? Bungenstock (2006) assumes that the southern North Sea is in the stage of a sedimentation coast, but this trend is defined in geological time scales.

The next chapter will deal with the new challenges people have to face resulting from climate change. A new approach in dealing with nature was introduced with the acknowledgment of the phenomenon of climate change that of sustainable development. The mutually consideration of social, economic and ecologic aspects for development. The Earth Summit in 1992 initiated a process called Agenda 21, and furthermore the concept of ICZM has been introduced. The growing awareness of nature conservation has led to a more comprehensive view on projects and programmes in the coastal zone. So, in consequence, coastal protection projects have been influenced by the interests and needs of nature conservation and this will continue.

What are the challenges imposed by climate change? What might the consequences be for coastal protection? The summary of the recommendations of the Advisory Committee for the North and Baltic Sea and the conclusions of Kunz (2004b) have led to the establishment of basic safety principles (p. 259):

- ... The existing philosophy of safety installed under the impression of the storm surges in 1953 and 1962 provides an overall concept for coastal protection which is still valid. The philosophy of safety is a sound basis for the enhancement of aims for coastal protection with regard to spatial protection concepts and strategies, which lead from reaction to adaptation and from solidification to flexibility.
- Coastal protection has to consider interests and needs which are derived from the principle of sustainable coastal development. [...] These cross-sectoral and multifunctional interests are asking for societal agreements about the desired safety in flood-prone areas. This stimulates a paradigm shift: From the principle of "security of failure for the dike line" to the principle that for certain areas a defined risk will not be exceeded (acceptance of risk) and that the remaining risks are manageable (risk management).

Chapter 3
New Insights – Varying Circumstances and New Frameworks

Contents

3.1 Climate Change and Consequences

Since 1988, the installation of the IPCC, four Assessment Reports have been published to estimate, to assess and to provide information about climate change and likely impacts. Although, initially the signals for climate change were not treated as significant, nowadays the existence of climate change is widely accepted (e.g. Climate Change Conference for post Kyoto time on Bali in 2007). The discourses and discussions focus mainly on the determination and the estimation of the repercussions on the climate system as shown in Fig. 3.1. The relationships and interactions are multi-faceted and complex and though it is somewhat difficult to predict the future development with a high level of certainty. Nevertheless, in the following section the main points of the recently published Assessment Report (AR4) shall be described.

F. Ahlhorn, *Long-Term Perspective in Coastal Zone Development*,
DOI 10.1007/978-3-642-01774-2_3, © Springer-Verlag Berlin Heidelberg 2009

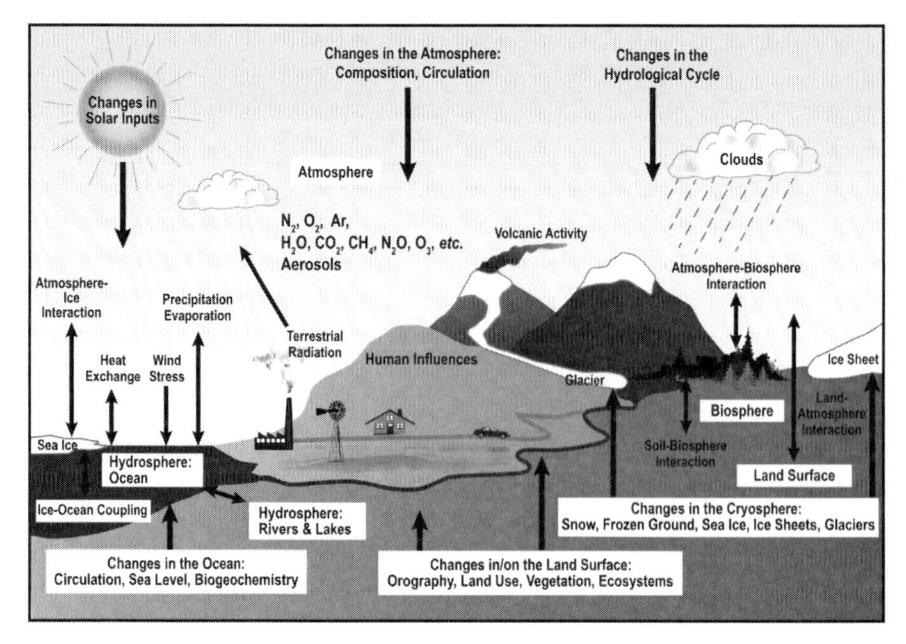

Fig. 3.1 Schematic view of the components of the climate system, their processes and interactions Source: Solomon et al. (2007), FAQ 1.2, Fig. 1.

3.1.1 IPCC – Fourth Assessment Report (AR4)

Recently, the IPCC released the Fourth Assessment Report (AR4) – see Solomon et al. (2007), Parry et al. (2007), Barker et al. (2007). The AR4 is divided into three groups: Working Group I (WG I) compiled and prepared the scientific basis, Working Group II (WG II) compiled a comprehensive overview of impacts, adaptation and vulnerability and Working Group III (WG III) prepared an exhaustive report on mitigation in respect of global climate change.

The WG I report is divided into four parts. The first part describes the changes of human and natural climatic drivers, the second part elaborates the observations of change in climate, the third part deals with attributing and understanding climate change and the last part describes the projection of the change into the future (Solomon et al. 2007).

The years 2005 and 1998 were the warmest years and the last 12 years were the warmest period since temperature recording in 1850. The increase of the average surface temperature is about $0.74^{\circ}C \pm 0.18^{\circ}C$ in the 100-year trend (1906–2005). The characteristics of global warming shows that the surface temperature over the land has increased faster than over the sea. The warming of the last three decades is spread over the earth, but the greatest warming was measured at higher northern latitudes (e.g. Hansen et al. 2006). Consequently, biological species have migrated polewards at 6 km per decade and vertically in the alpine regions at about 6 m per decade (Parmesan and Yohe 2003 in Hansen et al. 2006). "There is evidence of long-term changes in the large scale atmospheric circulation, such as poleward shift and strengthening of the westerly winds.... Many regional climate changes can be described in terms of preferred patterns of climate variability and therefore as changes in the occurrence of indices that characterise the strength and the phase of these patterns" (Solomon et al. 2007, p. 38). For example, for the occurrence and strength in the northern latitude the NAO index is crucial.

From 1961 to 2003 the sea level rise is estimated at 1.8 ± 0.5 mm per year. Since satellite measuring of sea level (1993–2003) the sea level rose about 3.1 ± 0.7 mm per year. If the latter rate reflects decadal variability of an accelerated increase of sea level the behaviour in the long run is uncertain. The rate of sea level rise has increased with *high confidence* from the 19th to the 20th century (Solomon et al. 2007).

"It is *extremely unlikely* (< 5%) that the global pattern of warming observed during the past half century can be explained without external forcing" (Solomon et al. 2007, p. 60). The conclusion is that "most of the observed increase in global average temperature since the mid-20th century is *very likely* due to the observed increase in anthropogenic greenhouse gas concentration.... It is *very likely* that the response to the anthropogenic forcing contributed to sea level rise during the latter half of the 20th century, but decadal variability in sea level rise remains poorly understood" (Solomon et al. 2007, p. 60).

For the mid-term projections of the average surface temperature and sea level rise is stated as: "Committed climate change[1] due to atmospheric composition in the year 2000 corresponds to a warming trend of about 0.1°C per decade over the next two decades, in the absence of large changes in volcanic or solar forcing. About twice as much warming (0.2°C per decade) would be expected if emissions were to fall within the range of the SRES marker scenarios.... Sea level is expected to continue to rise over the next several decades." (Solomon et al. 2007, p. 68).

According to the latest release of the IPCC, the sea level rise might range between 18 cm and 59 cm for the next 100 years (see Fig. 3.2). In Fig. 3.2 the model-based ranges are shown, which indicate different variations according to the underlying scenarios (see Sect. 4.3.1). The report of WG II (Parry et al. 2007) describes the observed impacts of the current climate change on biological and physical systems. The main findings are that for example "a global assessment of data since 1970 has shown it is likely that anthropogenic warming has had a discernible influence on many physical and biological systems [...], other effects of regional climate change on natural and human environments are emerging, although many are difficult to discern due to adaptation and non-climatic drivers" (Parry et al. 2007, p. 25). In general, WG II stated that more information about the effects of climate change on

[1] If the concentrations of greenhouse gases and aerosols were held fixed after a period of change, the climate system would continue to respond due to the inertia of the oceans and ice sheets and their long time scales for adjustments. *Committed climate change* is defined here as the further change in global mean temperature after atmospheric composition, and hence radiative forcing, is held constant. It also involves other aspects of climate system, in particular sea level (Solomon et al. 2007, p. 68).

Case	Temperature change (°C at 2090-2099 relative to 1980-1999)[a,d]		Sea level rise (m at 2090-2099 relative to 1980-1999)
	Best estimate	*Likely* range	Model-based range excluding future rapid dynamical changes in ice flow
Constant year 2000 concentrations[b]	0.6	0.3 – 0.9	Not available
B1 scenario	1.8	1.1 – 2.9	0.18 – 0.38
A1T scenario	2.4	1.4 – 3.8	0.20 – 0.45
B2 scenario	2.4	1.4 – 3.8	0.20 – 0.43
A1B scenario	2.8	1.7 – 4.4	0.21 – 0.48
A2 scenario	3.4	2.0 – 5.4	0.23 – 0.51
A1FI scenario	4.0	2.4 – 6.4	0.26 – 0.59

Notes:

a) Temperatures are assessed best estimates and *likely* uncertainty ranges from a hierarchy of models of varying complexity as well as observational constraints.

b) Year 2000 constant composition is derived from Atmosphere-Ocean General Circulation Models (AOGCMs) only.

c) All scenarios above are six SRES marker scenarios. Approximate CO_2-eq concentrations corresponding to the computed radiative forcing due to anthropogenic GHGs and aerosols in 2100 (see p. 823 of the Working Group I TAR) for the SRES B1, AIT, B2, A1B, A2 and A1FI illustrative marker scenarios are about 600, 700, 800, 850, 1250 and 1550ppm, respectively.

d) Temperature changes are expressed as the difference from the period 1980-1999. To express the change relative to the period 1850-1899 add 0.5°C.

Fig. 3.2 Projected global average surface warming and sea level rise at the end of the 21st century Source: IPCC (2007), Table SPM.1, p. 8.

physical, biological systems and human environments is available – e.g. Scholze et al. (2006). Impacts of climate change will vary regionally, but the costs will increase over time with increasing temperature. The costs of the consequences of climate change have been the issue of several publications – see e.g. Stern (2006) and Hübler et al. (2007). These findings and estimations of costs associated with climate change have inherent uncertainties, so that the figures only can indicate the range of costs: "This Review [Stern Review] has focused on the economics of risk and uncertainty, using a wide range of economic tools to tackle the challenges of a global problem with profound long-term implications. Much more work is required, by scientists and economists, to tackle the analytical challenges and resolve some of the uncertainties across a broad front. But it is already very clear that the economic risks of inaction in the face of climate change are very severe" (Stern 2006, p. 575). Adaptation strategies are available and some strategies have already been applied, but the extension of the level of adaptation needs to be extended. The vulnerability depends not only on climate change but also on other stresses. Sustainable development can reduce the vulnerability to climate change. Many impacts can be avoided, reduced or delayed by mitigation (Parry et al. 2007).

3.1.2 The European Dimension

The European dimension of impacts and consequences imposed by climate change are the topic of reports provided by the European Environment Agency (EEA): EEA (2004), EEA (2005) and EEA (2006).

The EEA report *Impacts of Europe's changing climate* (EEA 2004) describes the impacts of climate change according to a predefined list of indicators. These indicators are related to physical or biological systems as well as to certain sectors like agriculture, economic (in general) and human health. This report can be seen as a preparatory step for the technical reports of 2005 and 2006. For example, the trends described for the rising sea level are: "sea level around Europe increased by between 0.8 mm/year (Brest and Newlyn) and 3.0 mm/year (Narvik) in the past century, the projected rate of sea level rise between 1990 and 2100 is 2.2–4.4 times higher than the rate in the 20th century, and sea level is projected to continue to rise for centuries" (EEA 2004, p. 7). For the sector agriculture the report stated that the yields per hectare of all cash crops have been increased over the past 40 years due to technological progress. Agriculture might benefit from increasing CO_2 concentration and increasing temperature in the northern part, but this may lead to severe problems in the more southern part of Europe and harvests could deplete if more extreme events like droughts or floods occur in future.

The report on *Vulnerability and adaptation to climate change in Europe* (EEA 2005) provides a detailed description of the vulnerability of European countries and sectoral perspectives. The latter aspect is divided into two parts: the natural environment and associated services and the socio-economic sectors. The first aspect deals with, for example, natural ecosystems and biodiversity, agriculture, water resources and coastal zones. The observed increase in temperature and precipitation decrease already affect various components of natural ecosystems. Highly vulnerable regions have been identified, amongst them, several coastal zones across Europe. In 2050 approx. 80% of the 2000 current surveyed species would be lost and nearly 5% of the plants would lose their habitats according to the climate change scenarios (Schröter et al. 2005 in EEA 2005). For the agricultural sector the effects will be minor from a European perspective, but on national and regional levels respectively, there can be major impacts depending on the level of significance of this sector. Consequences are projected especially for the Mediterranean region. Depending on extreme events and on the weather conditions the sector can become highly vulnerable to climate change all over Europe. The impacts on the water resources in European countries strongly depend on the precipitation pattern and could exacerbate the situation. For tourism the snow cover is crucial for example for the alpine region, besides that heat waves can reduce tourism in certain regions, especially in the southern part of Europe. The coastal zones of Europe are vulnerable to sea level rise and the intensity and frequency of storm surges. The report stated that European policies and regulations take adaptation to the impacts of climate change into account, e.g. the Bird Directive and the Water Framework Directive (see Appendix A). As a national level research programmes have been initiated to investigate the implementation potential of adaptation strategies.[2] Finally, challenges on adaptation are described and suggestions for necessary research were made (EEA 2005, p. 8):

- Improving climate models and scenarios at detailed regional level, especially for extreme weather events, to reduce the high level of uncertainty.

[2] For example, on regional level in Lower Saxony the initiative of the Ministry of Science with KLIFF programme: Climate Impact Research Network. The main objectives are the extension of regional climate change models, to investigate the impacts to biological systems and to investigate and to develop regional or local adaptation strategies.

- Advancing understanding of "good practice" in adaptation measures through exchange and information sharing on feasibility, costs and benefits.
- Involving the public and private sector, and the general public at both local and national levels.
- Enhancing coordination and collaboration both within and between countries to ensure the coherence of adaptation measures with other policy objectives, and the allocation of appropriate resources.

The latter suggestions will be tackled within an Interreg project named "Climate Proof Areas", which involves different European countries such as The Netherlands, England, Sweden and Germany. The third report deals with *The changing faces of Europe's coastal areas* (EEA 2006). This report is a specific compilation of information about the trends and the state of coasts, current trends in policy responses and it proposes a conceptual framework for Europe's coastal zones. Major environmental problems in the North Sea are posed by the intrusion of foreign species and there are significant anthropogenic impacts. The North Sea coasts have high economic and population concentrations and are most vulnerable to coastal zone flooding (EEA 2006, p. 22). In Germany and The Netherlands more than 10% of the coastal area (0–10 km) has been designated as NATURA 2000 sites, less than 10% in the United Kingdom. The percentage for Germany and The Netherlands is much higher for the seaside area: 60 and 80% respectively (EEA 2006, p. 32). Recommendations made for a more sustainable development of Europe's coasts were as follows (EEA 2006, p. 82):

- Adaptive management of human values versus natural systems should be adopted.
- Priority should be given to sustaining and enhancing natural buffers, instead of depending on artificial coastal defences.
- Strategic land use planning should be implemented.
- Disaster preparedness should be considered through the development of relevant sections in the national climate change adaptation strategies and flood risk management systems.

3.1.3 Possible Effects of Climate Change in Germany

This section summarises the results of the regional climate change models for Germany based on the publications of Jonas et al. (2005) and Spekat et al. (2007).

In Germany climate change scenarios are developed on the basis of the REMO and WETTREG models using different approaches (see Sect. 4.3.1). REMO relies on the SRES emission scenarios provided by the IPCC (see Box on p. 105). The results of the REMO simulations are as follows:

- Until 2100 the increase of mean air temperature could vary between 2.5°C and 3.5°C depending on the further emission of green house gases.
- The increase of air temperature will vary regionally and seasonally.
- Until 2100 in winter time the mean air temperature might increase up to 4°C as compared with the time period between 1961 and 1990 (see Fig. 3.3).

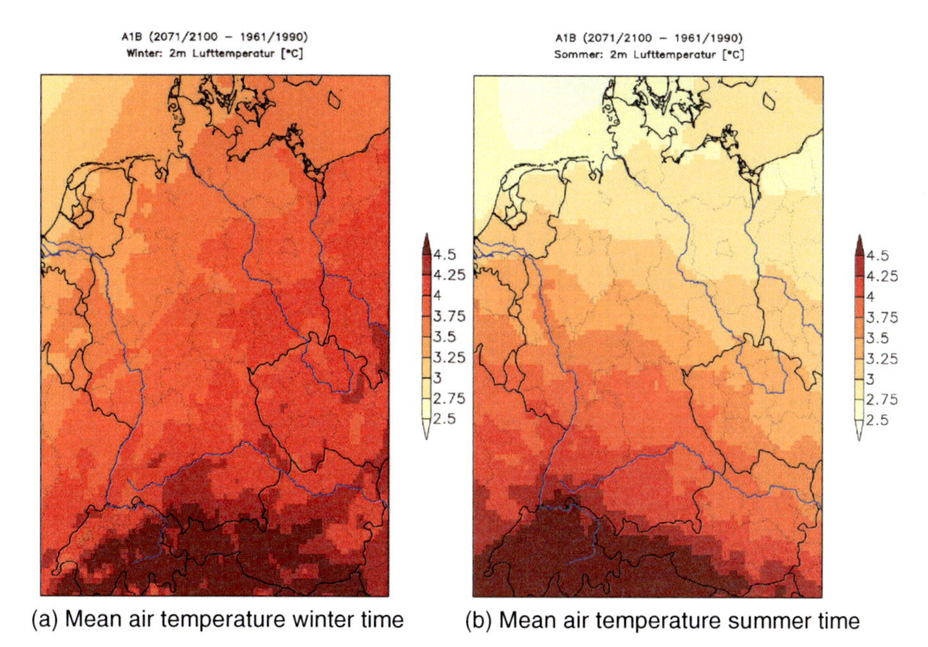

Fig. 3.3 Results of REMO for mean air temperature in winter and in summer time for 2071–2100 against the period 1961–1990
Source: MPI-M (2008).

- Precipitation might decrease in summer time and in some regions the decline of precipitation might be worse than in others.
- In winter time precipitation in some regions, especially in south and south-west Germany, might increase, but the occurrence of snow might decrease (see Fig. 3.4).

WETTREG also calculates the mean air temperature using the different scenarios for winter and for summer time in the period 2071 and 2100. The reference period is 1961–1990. In Fig. 3.5 the mean air temperature for summer time for the periods 2011–2040, 2041–2070 and 2071–2100 are shown. The results of WETTREG in respect of mean air temperatures are as follows (UBA 2007):

- The highest increase of mean temperature will occur in northern Germany and in the alpine region, with exception of the coastal region.
- The lowest increase of mean temperature in the North and Baltic Sea region and in the low mountain range.
- The mean temperature in Germany will vary according to the emission scenarios between 2.3°C (A1B) and 1.8°C (B1).
- The maximum calculated increase is approx. 2.5°C, and the minimum is approx. 1.5°C.

In Fig. 3.6 the result of the WETTREG simulation is shown displaying the development of the precipitation in winter time in the periods 2011–2040, 2041–2070 and 2071–2100. The precipitation might increase higher for the emission scenario

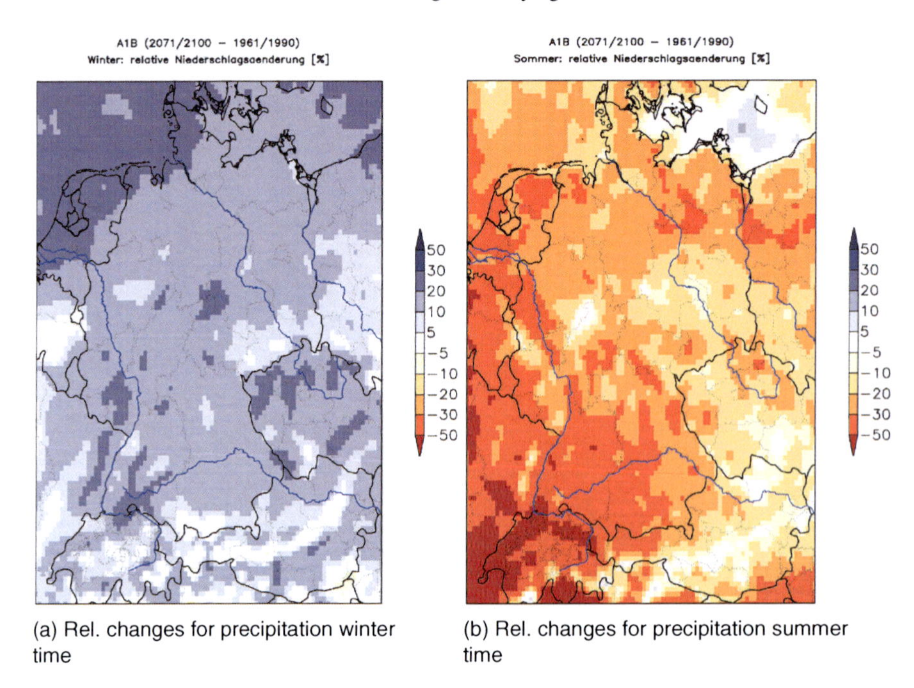

(a) Rel. changes for precipitation winter time

(b) Rel. changes for precipitation summer time

Fig. 3.4 Results of REMO of relative changes for precipitation in winter and in summer time for 2071–2100 against the period 1961–1990
Source: MPI-M (2008).

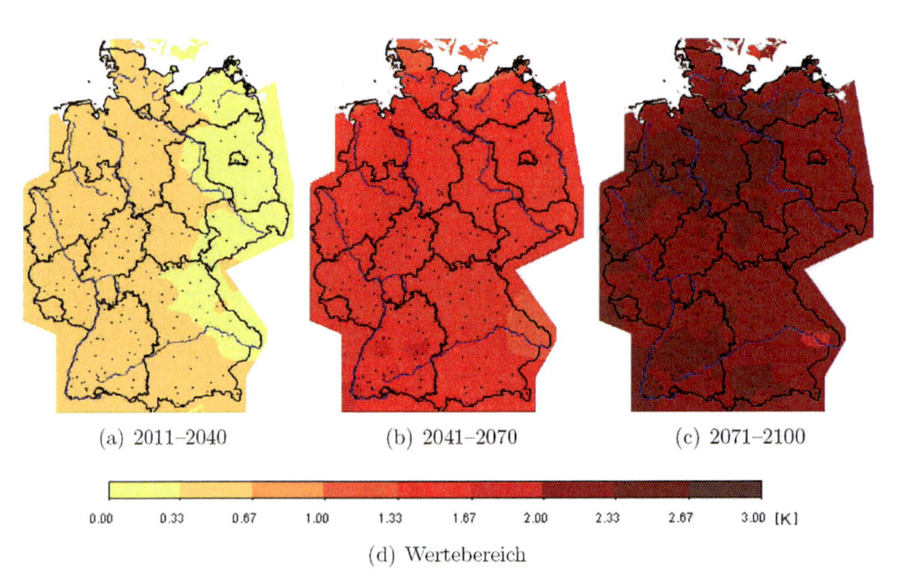

(a) 2011–2040

(b) 2041–2070

(c) 2071–2100

(d) Wertebereich

Fig. 3.5 Mean air temperature for summer time in three periods
Source: Spekat et al. (2007).

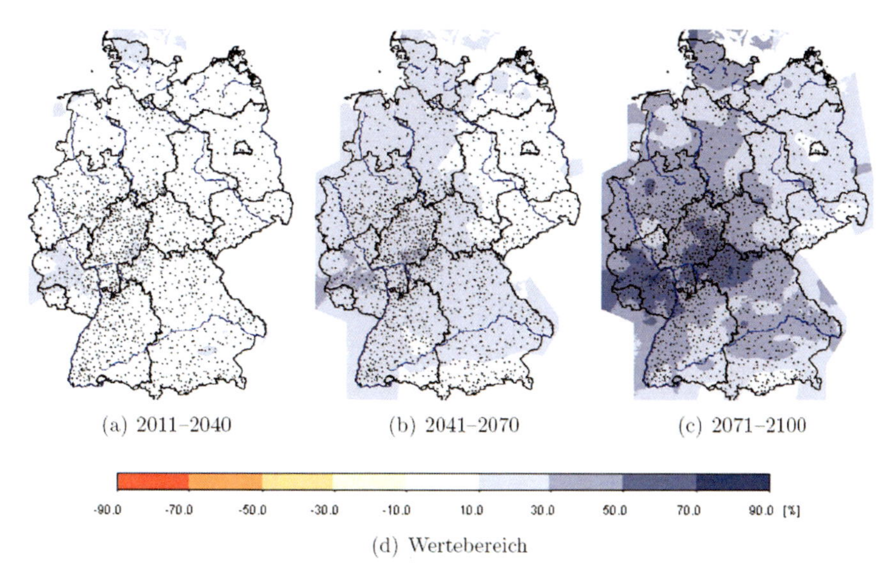

Fig. 3.6 Changes in precipitation in winter time for three periods
Source: Spekat et al. (2007).

A1B than for B1. The right-hand side of Fig 3.6 indicates that precipitation during winter time might increase higher in the western part of Germany than in the eastern part. For the North Sea coast WETTREG calculates an increase of about 60% for Schleswig-Holstein and the northwestern part of Lower Saxony. The simulations of the precipitation in summer time show a decrease in Germany up to 20%. The highest decline might be observed in the north-eastern part in Mecklenburg-Vorpommern. In other regions of Germany precipitation might decrease up to 10% (Spekat et al. 2007).

Furthermore, the results of WETTREG have been applied to the twelve different geographical regions in Germany. Important for this dissertation is the relevance for the coastal region and especially the north-western lowlands. The changes in climate relevant parameters will be presented in a semi-quantative way. For example, the results for the North Sea and Baltic Sea coast are compiled in Fig. 3.7. Temperature changes are featured at the top of the Figure: the light grey part of the diagramme on the left shows the control loop, the ⊖ symbol indicates that the temperature is cooler as compared with the mean for entire Germany. The next three parts of the diagramme (A1B, A2, B1) indicate the relative change of the temperature according to the emission scenarios relative to the changes for Germany as a whole. The change in temperature is similar to the mean of Germany as a whole. The three parts on the right-hand side indicate the changes in respect of summer days, hot summer days and tropical nights (definition depending on temperature). All items show an increase which is higher than the mean of Germany as a whole. The diagramme below the temperature indicates the changes in precipitation for the

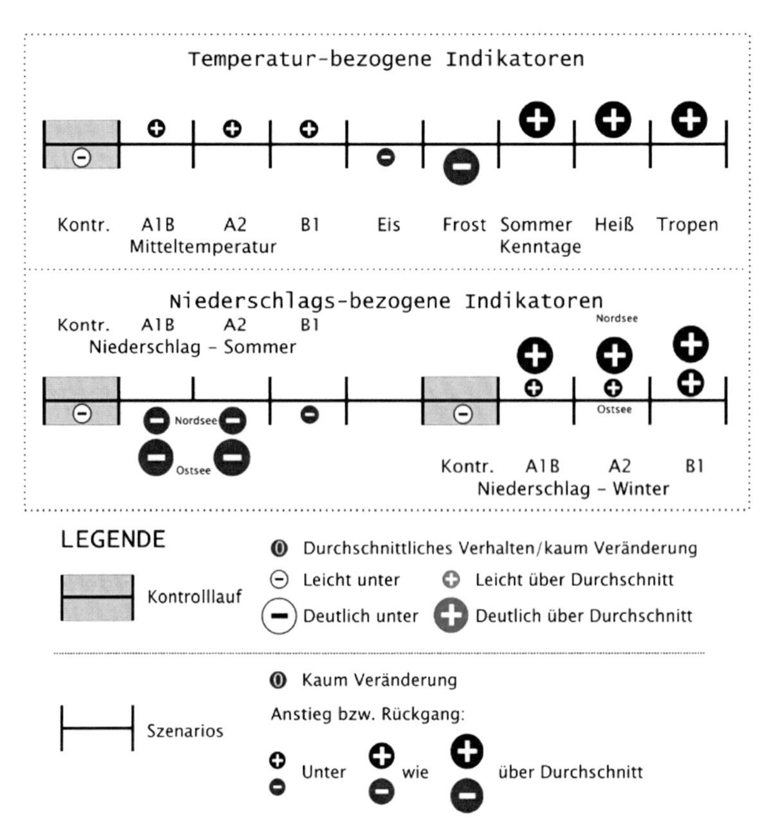

Fig. 3.7 Semi-quantative visualisation of the results of WETTREG for the North Sea and Baltic Sea coast
Source: Spekat et al. (2007).

North Sea and Baltic Sea coast. The symbols can be interpreted in the same way, differences between the North and Baltic Sea in precipitation are indicated.

Grossmann et al. (2007) stated that the mean maximum high water level at Cuxhaven at the end of the 21st century will range between 42 and 61 cm with a mean value of 50 cm. For 2030 the mean value for the mean maximum high water level for Cuxhaven might be 15 cm, with variation from 12 to 18 cm. The uncertainty for 2030 and 2085 taking additional effects into account, e.g. the unknown behaviour of the ice-sheets, amounts to ±20 cm resp. ±50 cm. The estimation of extreme storm surge events was dealt within the MUSE-research project. References and the main findings are summarised in the box on p. 47. In Fig 3.8 the scenarios of the changes in high water level are shown for the stations Cuxhaven and Hamburg St. Pauli. These scenarios are "based on the model TRIMGEO forced by winds and air pressure from different regional models and emission scenarios. Since A2 and B2 scenarios do not differ significantly, we indicate the mean value across all models and scenarios and the minimum maximum range" (Grossmann et al. 2007, p. 179).

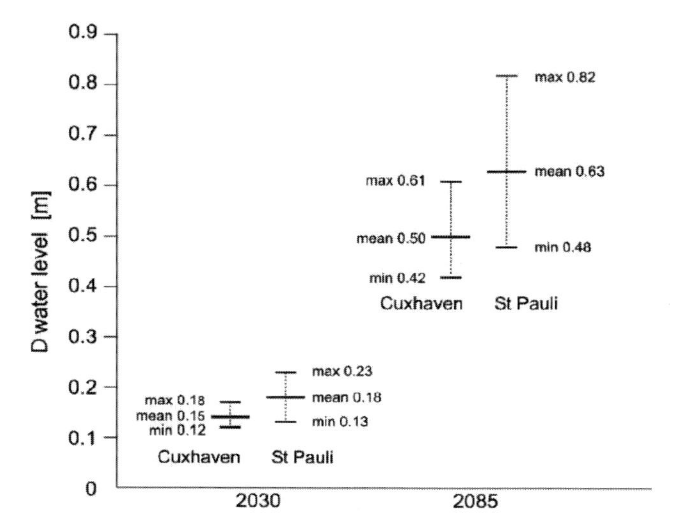

Fig. 3.8 Scenarios of changes in extreme high water level for Hamburg and Cuxhaven Source: Grossmann et al. (2007).

Matulla et al. (2007) examined storm behaviour and stated that the "storm climate in Europe has undergone considerable changes throughout the past 130 years and shows significant variations on a quasi-decadal time scale [...], the most recent years are characterised by a return to average or calm conditions [...] and the ability of the NAO index to explain storminess across Europe depends on the region and period under consideration" (pp. 5–6). The NAO index is able to explain the variations in storm behaviour over the last decades, but can not be used to explain the changes in former times (Matulla et al. 2007). von Storch and Weisse (2008) generally stated that until the end of the 21st century the westerly winds will intensify less than 10%. This will have slight effects on the storm surges and the wave heights in the southern North Sea.

MUSE – Investigation of Storm Surges with Very Low Probability of Occurrence

Another approach to estimate the probability of the possible height of extreme storm surges was carried out by the research project MUSE funded by the German Coastal Engineering Council (KFKI) and the Federal Ministry for Research (BMBF). The project was divided into three parts. In the first part the German National Meteorological Service (DWD) developed "physically extreme weather situations" and "extreme wind fields" which can possibly occur, but have not occurred yet (Koziar and Renner 2005). In the second part the Federal Maritime and Hydrographic Agency of Germany (BSH) used these extreme weather situations as input parameters for the ocean model to simulate storm surges (Bork and Müller-Navarra 2005). For example, the

results of the simulation of the modified 1976 storm surge showed that the highest water level observed in Cuxhaven (until June 2005) might be exceeded by up to 1.40 m. For Wilhelmshaven the highest water level observed was NN +4.87 m during the storm surge of 1976. This might be exceeded by approx. 1.6 m. The simulations show that due to these modified weather situations for 1976 the observed water levels along the German Bight might be exceeded by 1.0 m up to 1.6 m. These water levels are rated as likely to occur. The aim of the third part of the MUSE project was to determine the probability of occurrence. It was conducted by the University of Siegen (Jensen and Mudersbach 2005). For example, the probability of occurrence for the water levels of the modified weather situation in 1976 for Cuxhaven was calculated to approx. 10^{-4}. The highest observed water levels at the German Bight show a probability of occurrence of approx. 10^{-2}, the water levels with a probability of occurrence of 10^{-4} are about 60–110 cm higher.
To discuss the risk of flooding and to move from a deterministic to a probabilistic approach in coastal protection the knowledge of risks (e.g. likelihood of extreme water levels, of sea conditions, of failures of the defence system) is important. The MUSE project provides first insights in water levels which may, under certain weather conditions, occur. The boundary conditions have been predicted by using a specific simulation and forecasting technique based on existing data on extreme events like storm surges in the past. The weather conditions have been modified in a way that the occurrence of the new conditions are physically likely.

3.1.4 Regional Vulnerability and Sectoral Perspectives

Recently, Zebisch et al. (2005) published the results of an investigation on climate change in Germany – vulnerability and adaptation strategies of climate sensitive systems. The aim of this investigation was as follows:

- The documentation of the current knowledge about global climate change in Germany and the analysis of possible impacts of climate change of seven sectors (Tourism, Water Management, Agriculture, Forest Management, Biodiversity/Nature Conservation, Health and Traffic).
- The investigation of the current degree of adaptation and the capacity for adaptation of several climate-sensitive sectors.
- The determination of the vulnerability of distinct sectors and regions in Germany through the comparison of the above mentioned aspects.

The results of the study have been discussed with decision makers in government, politics and the economy to develop a basis for an adaptation strategy in Germany. In Fig. 3.9 an excerpt of the main findings are shown. The red colour indicates *high*, the orange colour *medium* and the yellow colour *low vulnerability*, respectively. For the coastal zone a *high vulnerability* is indicated for high water, i.e. storm surges and sea level rise. The sectors agriculture, forestry and health display a low vulnerability. Biodiversity and nature conservation are classified as *medium vulnerability*, but this classification is subject to uncertainties. The classification of the vulnerability of biodiversity and nature conservation is split up into "protection of the status quo" and "protection of processes". For tourism *medium vulnerability* was investigated. In total, Zebisch et al. (2005) classified the vulnerability of the German coastal zone and the low-lying north-western areas as *medium*.

The research project "Climate Change and the Coast – The Future of the low-lying marshes of the river Weser (KLIMU)" (Schuchardt and Schirmer 2005b) investigated the consequences of climate change for the north-western part of Lower Saxony. Within this project the human-environment interaction was reflected by three networks: the region Weser-estuary, coastal protection, agriculture. The impacts of climate change, set by certain scenarios for the pilot area, were investigated within several sub-projects. Each of these networks has been divided into three parts according to the instruments needed to steer and to manage the network, the natural space and the socio-economic sector. The results reflect both the primary and the secondary impacts on each network and its parts. For example, the primary impacts of climate change for the *Weser-estuary* network are related to the hydrography, the water levels, moisture of the soil, water quality, the biotop-types and agriculture. Hence, the given sea level rise of 70 cm will lead to an increase of

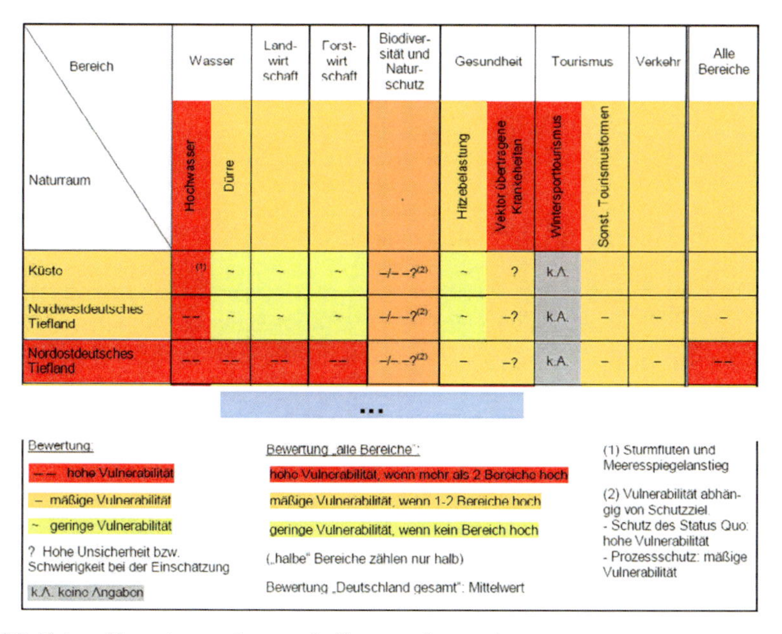

Fig. 3.9 Vulnerable regions and sectors in Germany (excerpt)
Source: Zebisch et al. (2005).

the tidal range in the city of Bremen. For the sector agriculture the impacts may lead to a limited utilisation of the fore land in front of the main dikes. Primary impacts on the *coastal protection* network are related to the technical consequences for the *Weser-estuary*. For example, different strategies have been investigated to adapt to the anticipated sea level rise. The existing coastal protection system has been advanced, for example, with a barrier in the mouth of the Weser or the necessary measures to adapt the dikes to the higher water levels (Zimmermann et al. 2005). The safety of the main dikes along the river Weser was calculated according to the probability of exceedance of the design water level. Zimmermann et al. (2005) assumed the exceedance of the design water level and thus the overtopping of the main dike as a failure mechanism. Consequently, the probability of occurrence of certain wind and wave levels has been used to calculate the wave height on certain coastal stretches along the river Weser and the Jade Bay. Finally, the probability of recurrence of a certain wave run-up has been calculated under present conditions and under the new conditions given by the scenarios. For example, the probability of recurrence of 1–1,000 for present conditions will be reduced to 1–200 if the sea level rise is 70 cm and the wind speed increases up to 3.8% (as given in the KLIMU scenarios). For the *agriculture* network the KLIMU climate change scenario leads to increasing profits due to an increase in temperature and CO_2 concentration, the effects of higher precipitation and higher water levels in the river Weser. The results showed, that the calculated impacts can be handled by the existing water management system (Schuchardt and Schirmer 2005a).

3.2 Sustainable Development and Integrated Management

The principle of sustainable development is widely known and has been widely discussed and treated in the literature (UN 1992). Started in the early 1970s within the emerging nature conservation movement, it entered the political arena in 1992 at the Earth Summit in Rio de Janeiro, prepared by Brundtland's Commission (WCED 1987).

An interesting approach to discuss the interlinkage between sustainable development and ICZM was given by Daschkeit (2004). Daschkeit discusses comprehensively the problems and challenges which might occur in tackling sustainable development and applying this principle to the coastal zone, i.e. to implement ICZM processes. The main points which were discussed are: (a) problems of knowledge and activities in achieving sustainable development, (b) integration, (c) globalisation versus regionalisation, (d) "round-table"-syndrome, i.e. participation. Here, the key points will be discussed against the background of the approach described in this dissertation.

First: The problems of knowledge and activities needed to achieve sustainable development relate to the relationship between politics and science. After launching 'sustainable development' onto the political agenda, the key question was "What decisions do we have to take to achieve sustainable development?" Thus, a new branch of research was created and the question was transformed into: "What do we have to know to achieve sustainable development? and What kind of knowledge do we need?" (Daschkeit 2004). Approximately 20 years after Rio the debate is still clouded by diverging opinions, approaches and methods which promise that their application will lead to sustainable development. But, how can sustainable development be defined? The basis for sustainable development is the insight that at least three pillars have to be considered: social, economical and ecological aspects. This problems about knowledge and action seems to resemble the situation of coastal protection in Germany. Considerable knowledge about the consequences of climate change and the likely impacts are available, but the main actors of coastal protection insist on the strategy of "taking action only underwritten by secure knowledge" (Lange et al. 2007). This is understandable in the light of the good experiences over the last decades, but the question whether "hold the single line" and construct the dikes applying the empirical approach of "single value procedure" is the right strategy? Important options have already been propagated over the last years, e.g. strengthening or building of additional embankments in the hinterland. This is a question of risks/uncertainty versus safety. That means, that knowledge of the evolution of climate change is per se uncertain and can only be estimated as conscientiously as possible, but the knowledge needed to build e.g. a dike should be well-founded. However, the failure mechanisms of dikes are poorly understood and are the focus of current research. This will be discussed in detail in Sect. 3.3 of this chapter.

Second: The difficulties of the integration between the three pillars, the social, economical and ecological aspects. The integration of different sectors is complicated enough and has been widely treated in several projects on sustainable development. Furthermore, problems also arise related to the three-pillar integration within a sector. Daschkeit (2004) mentioned that old industrial areas are seen as counterproductive in terms of sustainable development, but if the areas were to be populated by rare animals or plants, they would be regarded as highly valuable. Cooperation can be difficult even if the actors pursue similar goals. Conflicting interests become evident e.g. for the salt marshes (fore land) in front of a main dike: nature conservation issues call for an undisturbed and large salt marsh and also coastal protection issues want a broad fore land to protect the main dike foot. But, there is a big dispute about the maintenance of the fore land. Cooperation issues also features in the preparation of plans and programmes. Nowadays, these plans and programmes are prepared by sectors with a narrowed sectoral view. Real consideration of other sectors is sparse. And if consideration takes place, it is mainly caused by formal procedures.

Third: The problem between globalisation and regionalisation will not discussed further, because this problem stretches beyond the objectives of this dissertation.

Fourth: The "round-table"-syndrome has a direct relationship to this dissertation. According to Daschkeit (2004) the decision-making processes are characterised by a sense of weariness and a lack of enthusiasm. It is true that participation has increased over the last decades, which has been stimulated by the developments of instruments, like the Århus Convention (UN/ECE 1998) on international level. This convention stresses active involvement and free access to information for the general public. But, this has not penetrated all the instruments and rules in force. If this is desirable or not and what amount of participation is effective or not has still to be answered. However, this is the objective of a big research branch called "governance": "Governance is the sum of the many ways individuals and institutions, public and private, manage their common affairs. It is a continuing process through which conflicting or diverse interests may be accommodated and co-operative action may be taken. It includes formal institutions and regimes empowered to enforce compliance, as well as informal arrangements that people and institutions either have agreed to or perceive to be in their interest" (Source: www.libertymatters.org/chap1.htm, May 2008). The demand for participation is still present and will be stimulated by current development of tools and instruments. But, are these instruments adequate?

This dissertation is based on a Participatory Integrated Assessment (PIA) process conducted at the coastal zone to achieve sustainable development. No matter where the 'governance' debate leads to one key prerequisite of sustainable development would be the establishment of a joint agreement on necessary and adequate action to anticipate and to adapt to future development.

3.2.1 The International Perspective

The intention of this section is not to add another definition of the term Integrated Coastal Zone Management to the vast amount of publications about ICZM. In the Baseline 2000 report Sorensen (2000) compiled many documents, reports and projects dealing with Integrated Coastal Management (ICM). The beginning of integrated thinking in planning and management of the coast is located in the mid 1960s. From there on, many steps were taken on the way to the Earth Summit, where ICM is the main topic in Chap. 17 of the UN document: Protection of the oceans, all kinds of seas, including enclosed and semi-enclosed seas, and coastal areas and the protection, rational use and development of their living resources (UN 1992). Sorensen (2000) created a differentiation between ICM and ICZM: "ICZM requires that the planning and management must include a zone comprised of: (1) coastal and estuarine waters, (2) the adjoining and complete inter-tidal area, and (3) the supra-tidal coastal lands. The coastal lands should extend inland to at least the maximum highest tide and include directly connected coastal environments such as wetlands and dune systems" (p. 3). Taking this definition of the area of validity into account the term ICZM is most appropriate for this dissertation.

Nevertheless, to understand the term ICZM this dissertation draws on the definition of The World Bank given in Post and Lundin (1996, p. 1): "ICZM is a process of governance and consists of the legal and institutional framework necessary to ensure that development and management plans for coastal zones are integrated with environmental (including social) goals and are made with the participation of those affected. The purpose of ICZM is to maximise the benefits provided by the coastal zone and to minimise the conflicts and harmful effects of activities upon each other, on resources and on the environment". The intention of this dissertation is to offer methods and tools to meet this definition as far as possible. Furthermore, ICZM is a dynamic process and pro-active rather than re-active, but as a process it can also function on re-active basis as well.

Obviously, with the beginning of ICZM a new way of thinking was required. The traditional sectoral approach of planning and management has to be substituted by integrated planning and management. The areas of jurisdiction of plans and programmes have to be revised. And the re-active and mainly static approach of spatial planning has to be transformed into a pro-active and dynamic process. This requires adequate and appropriate methods, tools and instruments for both planning and management. Within the last decades numerous of research projects dealt with these issues and developed several methods and tools. However, whereas research has provided tools and methods, the incorporation and implementation in planning processes and programmes is often to slow. For example, it took about 10 years after the Earth Summit for the European Commission to publish the EU recommendations on ICZM (EC 2002), and the Member States need far more time to develop and implement national ICZM strategies.

3.2.2 The European Perspective

The EU initiated a demonstration programme on ICZM from 1997 till 1999 – see e.g. EU (1999). More than 30 projects along the European coast dealt with the issue of ICZM. The aim of these projects was to identify the potential of and the obstacles facing the implementation of ICZM in European countries.

Only one project concentrating on ICZM and spatial planning will be discussed in more detail: the NORCOAST project (NORCOAST Project Secretariat 2000). This EU Interreg project exercised a strong influence on the development of the ICZM recommendations by the EU. The project aims were to investigate and to promote good practice in ICZM. NORCOAST was based on the experiences and knowledge of practitioners in spatial planning and management on a regional level throughout Europe. Thus, the recommendations of the NORCOAST project focus mainly on the integration of spatial planning into ICZM and what this entails. These recommendations distinguish between the process of planning, the planning techniques and the regulatory framework (see box below). Some of these recommendations have been taken into account and tackled in the ComCoast project, especially in the German case study. For the process of planning it is crucial to involve and engage the public in an early stage and to initiate an integrated approach of solving a problem.

NORCOAST developed its recommendations on the basis of existing problems at the coasts around the North Sea. All relevant types of land use were taken into account. Recommendations were also made for specific issues like sea level rise and coastal protection. For these, the recommendations are as follows:

- Develop sustainable regional and national strategies for coastal protection based on dynamic conservation approaches making use of and working with the natural processes as far as possible.
- Develop appropriate and integrated coastal process research, information and monitoring to guide future planning and management.
- Identify areas which may be subject to future flooding and erosion as a result of sea level rise and take these into account in developing spatial planning policies.
- Financial provision for coastal protection should allow for compensation for planned loss of property and land in managed retreat scenarios.
- Use coastal sediment cells as the basis for coastal protection planning and management but ensure this is integrated within the system for comprehensive spatial planning.
- Provide integrated national and regional strategies for coastal protection planning with adequate national funding in support of regional works.

General Recommendations of the NORCOAST Project

- Process of Planning
  - ✓ Aim for an integrated approach to reduce conflicts and build synergy
  - ✓ Involve all relevant stakeholders and politicians
  - ✓ Make the process transparent, accountable, open and consultative
  - ✓ Identify a lead agency to initiate and facilitate the ICZM process
  - ✓ Develop a clear vision for the coastal zone
  - ✓ Establish coastal fora or partnerships to develop a shared sense of stewardship
- Planning Techniques
  - ✓ Describe possible consequences
  - ✓ Consider different scenarios
  - ✓ Accept that the coastal area is an open system
  - ✓ Aim for flexible planning
- Regulatory Framework
  - ✓ Legislate for a clear statutory responsibility for spatial planning - land and sea
  - ✓ Appoint authorities as lead agencies to initiate ICZM
  - ✓ Define a national framework for ICZM
  - ✓ EU should provide practical support for the development of ICZM

Other European Interreg projects like EUROSION, for example, tried to act on these recommendations. EUROSION dealt with the process of erosion along the entire European coast and made recommendations on the treatment of coastal erosion in the future – see e.g. EUROSION (2004a). Within the ComCoast project a brief investigation was made of the follow-up actions of NORCOAST in the ComCoast partner countries – see Ahlhorn et al. (2006). The result of this investigation was that no real follow-up actions were effected. The project was recognised as a well done project with good recommendations, but totally lacking in essential influence. The changes and adaptations made in partner countries which met the NORCOAST recommendations were, however, initiated for different reasons. For example, the extension of the State Spatial Planning Programme of Lower Saxony (LROP) to the 12 nm zone was enacted on the request of a near-shore wind farm project developer. The same is valid for The Netherlands, the UK and Denmark. In The Netherlands and the UK changes and adaptations of the spatial planning system are imminent, but these do not directly refer to the NORCOAST project. A first integrated approach in Lower Saxony for the coastal zone (land and sea) is the Spatial Concept for the Coast of Lower Saxony (ROKK: Raumordnungskonzept für das niedersächsische Küstenmeer, ML-RVOL 2005, see Sect. 4.1).

The outcomes and results of the demonstration programme led to the EU recommendations on ICZM (2002/413/EC) (EC 2002). The recommendations on ICZM comprise the strategic approach, the principles, the national strategies and the cooperation on ICZM. The ICZM principles crucial for the process are as follows:

- a broad overall perspective,
- a long-term perspective,
- adaptive management,
- specific solutions and flexible measures,
- working with natural processes and respecting the carrying capacity of ecosystems,
- involve all the parties concerned,
- support and involve the relevant administrative bodies at national, regional and local level and
- use a combination of instruments.

These principles are the fundamental basis for all ICZM endeavors. The challenge is to fill these principles with real and practical life. Many projects have been launched to fulfil and to enhance the vitality of these principles. For example, in Germany two projects, *Coastal Futures* and *ICZM Oder*, aim to investigate the opportunities and challenges of the implementation of ICZM principles. *Coastal Futures* deal with the aspect of off-shore wind farming and the related opportunities and challenges both at sea and on land (Colijn and Kannen 2003), and *ICZM Oder* aims to investigate the opportunities and challenges of integrated management for a river catchment area in cooperation with Poland – see e.g. Feilbach (2004), Schernewski et al. (2007).

Furthermore, the recommendations on ICZM included the development of national strategies by 2006. Almost all European countries have documented their national stocktaking and strategy: for The Netherlands see RWS (2005) and for Germany see BMU (2006). Each country has adopted its own approach to ICZM. In August

2006 the evaluation of the national strategies was published. The main findings are as follows (Rupprecht Consult 2006, pp. 219–220):

- Integrated approaches to manage the interests in the coastal zone have been scarcely implemented and were not strategically employed, except on a case study basis.
- Due to the particular historic contexts of EU countries in their planning procedures and processes, the range of measures on dealing with the coastal zone is extremely diversified.
- The potentials of the current EU ICZM recommendations have not yet been fully exploited; an incentive-based approach will be more effective on the European level.

These findings demonstrate that the EU principles on ICZM have not been fully implemented, only on a case study level. Consequently, it can be stated that the principles are sufficient and necessary for the current stage of ICZM in European countries. If all projects under the umbrella of ICZM strive to reach these principles or adopt these principles as guiding principles for their own approach, then on the long-term the EU can reach the declared target of integrated planning and management in coastal zones. Projects taking these principles into account, like the ComCoast project, can take the process a step further. The specific historical, cultural and social development of EU countries can hamper the implementation of a common strategy on ICZM. Real cooperation and extensive exchange of knowledge between EU countries on different levels in planning and management can lead to a better management of the European coasts. And, perhaps it is not wise to implement a common strategy as long as the basic principles are viewed solely as guidance.

3.2.3 National Strategy on ICZM in Germany

The national strategy on ICZM for Germany (BMU 2006) is both a stocktaking exercise and an approach of developing a strategy on ICZM. The stocktaking element describes the present situation and distinguishes between economic stakeholders, additional stakeholders, activities and instruments in the German coastal zone. The stocktake of each aspect is divided into four parts: status-quo, perspectives and strategies, legal framework, economical, ecological and social relevance and conflicts. Coastal protection and spatial planning are separately discussed in the stocktake.

The second part of the national ICZM strategy in Germany deals with the "next steps towards a national strategy". Four pillars of a national ICZM strategy are outlined: (a) optimisation of statutory instruments related to the ICZM principles, (b) provision of the conditions for the continuation of the dialogue process, (c) best practice projects and their evaluation and (d) development and application of ICZM indicators. Important for this dissertation is the analysis of the strengths and weaknesses of the existing instruments. The available instruments will be investigated in the light of four principles: (1) sustainable development, (2) integration, (3) participation and communication and (4) exchange of knowledge.

The review of the first principle "sustainable development" is focussed on the consideration of ecological aspects within existing instruments. Furthermore, little is said about what sustainable development could include within coastal zones, although the principles of ICZM had been described previously. The second principle "integration" concerns the integration of land and sea and of institutions. Options for good integration were identified in formal planning approval processes. The identified weaknesses are the sectoral differentiation of responsibilities for planning issues and a lack of interaction between different sectors. The strengths related to the third principle "participation and communication" were identified as the formal participation procedures within both planning approval processes and the installation of the environmental information law on Federal and State level. The weakness of this principle is the insufficient communication between different actors in coastal zones and the varying application of participation procedures. The review of the "exchange of knowledge" – principle identified as a strength the comprehensive monitoring data and information on spatial issues, and the means of better accessing these data. The weakness in the implementation of this principle was the insufficient opportunity to demonstrate the achievements and the status of the sustainable development process, which to some extent was not possible on the basis of available knowledge, information and data. Practical suggestions for the continuation of the dialogue were the installation of an ICZM secretariat and the establishment of coastal fora for the North and the Baltic Sea. Suggestions have been made for the improvement of instruments and approaches currently available in Germany.

3.3 Addressing (Flood) Risk Management

3.3.1 Risk, Uncertainty and Vulnerability

"All decisions are intended to bring about some future benefit to someone or something, and involve choices (e.g. whether to act, to implement policy A or B, etc.). Without uncertainty, these decisions would be straightforward. Reality, however, is far more complex and hence all decisions involve judgments regarding uncertainty, understanding how they contribute to decision uncertainty, and the management of uncertainties within the assessment and decision-making process, are therefore essential to making well-informed decisions. While not all decisions produce the benefits that were intended, any decision should, even with the advantage of hindsight, be justifiable on the basis of the available knowledge at the time of the decision" (UKCIP 2003, p. 43). Thus, almost all decisions are made with uncertainty, and that is also valid for the case of coastal protection. The term uncertainty can be further divided into (a) knowledge uncertainty, (b) natural variability and (c) decision uncertainty (Gouldby and Samuels 2005). The background of "knowledge uncertainty" is clearly defined, it arises from the lack of knowledge about the behaviour of the investigated system. "Natural variability" encompasses the inherent variability of the real world. The "decision uncertainty" reflects the complexity of the social and organisational values and objectives, and the consequences decisions may have in the future.

For example, man decided to settle in the flood-prone area and experienced threats e.g. by storm surges or inundations of high fluvial water. The decision to protect themselves against flooding and the devastating consequences were made under uncertainty, especially, a long time ago when "engineering" was based on intuition and accumulated experiences. Today's evaluation of these decisions has to consider the formerly available knowledge and the circumstances facing the coastal community. Nowadays, the advantage of forecasting lies in lowering the probability of negative future consequences of decisions. Therefore, dealing with risk is twofold (a) current risk management of the status-quo with the ingredients such as analysis, evaluation and management and (b) taking the consequences of changes in climate, demography and economy into account as early and as accurately as possible. In this section some general remarks will be given on the theoretical background, but the main focus will be made on the relationship to coastal protection. For that reason, this section is mainly based on the outcomes of Markau's dissertation. The dissertation develops an approach for the risk analysis, evaluation and management for natural hazards exemplified in the flood-prone area of the State Schleswig-Holstein.

Markau (2003) stated that no consistent definition existed for the term "risk" and the methods to deal with risk. A European Flood Risk Management project (Floodsite) collated in its first phase information about the language of risk in different partner countries (Gouldby and Samuels 2005). The following definitions are given in UKCIP (2003) and Gouldby and Samuels (2005):

Risk is the combination of the probability of a consequence and its magnitude. Therefore risk considers the frequency or likelihood of occurrence of certain states or events (often termed 'hazards') and the magnitude of the likely consequences associated with those exposed to these hazardous states or events.

Uncertainty exists where there is a lack of knowledge concerning outcomes. Uncertainty may result from an imprecise knowledge of the risk, i.e. where the probabilities and magnitude of either the hazard and/or their associated consequences are uncertain. Even when there is a precise knowledge of these components there is still uncertainty because outcomes are determined probabilistically.

Risk analysis is a methodology to objectively determine risk by analysing and combining probabilities and consequences.

Risk perception is the view of risk held by a person or group and reflects cultural and personal values, as well as experience.

Risk management is the complete process of risk analysis, risk assessment, option appraisal and implementation of risk management measures.

Risk communication is any intentional exchange of information on environmental and/or health risks between interested parties.

Vulnerability is the characteristic of a system that describes its potential to be harmed. This can be considered as a combination of susceptibility and value.

The same is valid for the determination of risk, for which different formulas are available. In technical projects risk will be determined by the product of probability of failure and vulnerability (Gouldby and Samuels 2005). This approach comprises advantages as well as disadvantages, the detailed elaboration will exceed the scope of this dissertation and can e.g. be found in Markau (2003), Gouldby and Samuels (2005) or Kaiser (2006). For practical reasons and the purpose of this dissertation the following definition will be adopted:

$$\text{Risk} = \text{Probability of Failure} \times \text{Vulnerability} \quad (3.1)$$

Risk can be tackled by influencing two items: the probability of failure or the vulnerability. The German Advisory Council on Global Change (WBGU) proposes a system of classifying risks (WBGU 2000), which is based on the following criteria (p. 58):

- Probability of occurrence
- Magnitude of damage
- Ubiquity (spatial dimension of damage)
- Persistence (contaminants accumulated over long periods)
- Irreversibility (damage can not be remedied)
- Delay effect (time period between the event and the damage)
- Mobilisation (risks lead to severe conflicts and dread among the general public)

The WBGU differentiates between eight types of risk (e.g. damokles, cyclops), and suggest a decision tree to classify these types. Three areas were identified for the different types of risk: (a) normal, (b) transitional and (c) prohibited area (see Fig. 3.10). The classification of these areas is determined by the probability of occurrence and the extent of the damage. Markau (2003) concludes after the determination and evaluation of the risk of flooding, that storm surges are of the cyclops type and hence are located on the border between the transitional and prohibited area. For each type of risk defined by the WBGU a strategy or at least a minimum of necessary action is required and proposed.

For example, the cyclops risk type demands strategies as shown in Fig. 3.11. Three strategies have been identified to conduct the cyclops risk type: (a) ascertain the probability of occurrence, (b) prevent surprises and (c) emergency management. Concerning the first strategy the need to determine or to assess the probability of occurrence based on research or technical measures is emphasised. Within the second strategy technical measures have to be taken to prevent surprises, and for the third it is recommended that the individual and institutional capacities be strengthened and technical measures to restrict the damages be taken.

Another essential component is the knowledge and evaluation of the vulnerability of coastal zones, which refers directly to the second term of the risk formula. For example, various aspects of vulnerability to climate change impacts are outlined in Klein and Nicholls (1999). An overview on state-of-the-art for regional and local vulnerability assessment can be found in Sterr et al. (2000). The "overview" proposes an integrated process to identify the most appropriate coastal adaptation

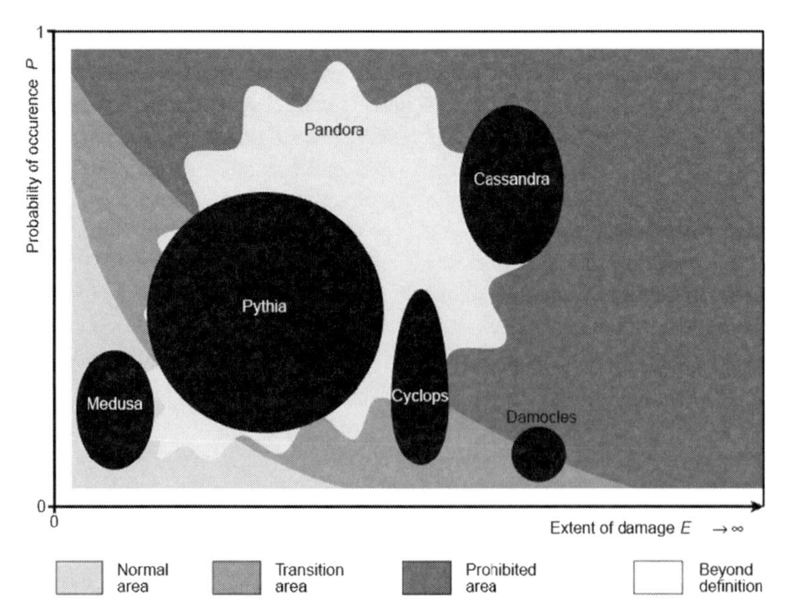

Fig. 3.10 Classification of risk after WBGU
Source: WBGU (2000).

Strategies	Tools
1. Ascertaining the probability of occurrence *P*	• Research to ascertain numerical probability *P* • International monitoring through – National risk centers – Institutional networking – International Risk Assessment Panel • Technological measures aimed at estimating probabilities
2. Preventing surprises	• Strict liability • Compulsory insurance for risk generators (e.g. floods, settlements) • Capacity building (licensing procedures, monitoring, training etc.) • Technological measures • International monitoring
3. Emergency management	• Human-resource and institutional capacity building (emergency prevention, preparedness and response) • Education, training, empowerment • Technological protective measures, including containment strategies • International emergency groups (e.g. fire services, radiation protection etc.)

Fig. 3.11 Strategies and instruments for the risk type cyclops
Source: WBGU (2000).

strategy. Sterr et al. (2000) evaluated two approaches to assess the vulnerability of coastal regions: the IPCC Common Methodology for Assessing the Vulnerability of Coastal Areas to Sea-Level Rise (result of the application for Germany e.g. Ebenhöh et al. 1996 and for The Netherlands the ISOS-study by Peerbolte et al. 1991) and the methodology developed by UNEP which is based on the IPCC methodology. Several disadvantages have been identified, such as shortage of accurate and complete data or that the methodology is "less effective in assessing the wide range of technical, institutional, economic and cultural elements present in different localities" (Sterr et al. 2000, p. 5). Therefore, a conceptual framework for coastal vulnerability assessment was suggested to overcome the deficiency of the existing methodologies (Fig. 3.12). The framework is divided into the socio-economic and the natural system, since the assessment has to incorporate both sides of the coin. Climate change and the consequential sea level rise impose effects on both systems, and both systems have different means of reacting or adapting to changing circumstances. The deficiency of the Common Methodology was to focus on the impacts of an accelerated sea level rise and its consequences for coastal zones. Sterr et al. (2000) concluded that "coastal adaptation requires data and information on coastal characteristics and dynamics, patterns of human behaviour, as well as an understanding of the potential consequences of climate change. It is also essential that there is a general awareness amongst the public and coastal planners and managers of these consequences and of the possible need to act" (p. 14).

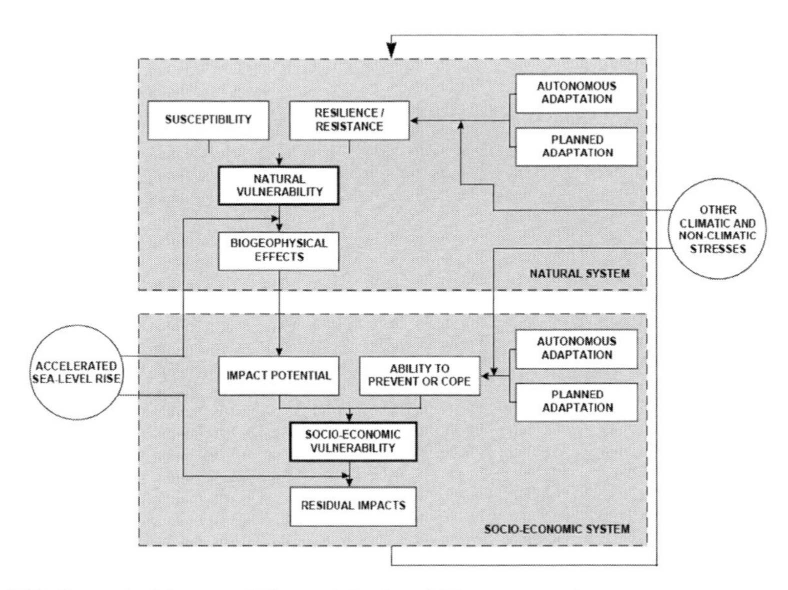

Fig. 3.12 Conceptual framework for coastal vulnerability assessment
Source: Sterr et al. (2000).

3.3.2 Risk Perception

"The day-to-day, 'intuitive' perception and evaluation of risks (in short: risk perception) is a basis for acting and behaving in dangerous situations. It is also fundamental for decisions such as whether preventive protective measurements are taken or not" (Plapp 2001, p. 2). Formerly, risk analysis and risk management based mainly on the technical or scientific approaches and was defined by the risk of technical constructions like nuclear power plants. The same is valid for coastal protection – e.g. Giszas (2003), Kunz (2004b), RIKZ (2002) for The Netherlands. "Scientists and decision-makers who assess and develop strategies to reduce the vulnerability of coastal zones to natural hazards and to improve disaster preparedness have to consider aspects in addition to quantitative measurable determinants such as inundation depth or the number of people affected. Whether a hazard has negative impacts on coastal societies or turns into a disaster depends to a great extent on human behaviour. The human behaviour in turn depends not on facts, but on perception, experience and knowledge" (Kaiser 2006, p. 158). Plapp (2003) carried out a comprehensive empirical investigation of risk perception of natural hazards (storm, high fluvial water, earthquakes) in different areas of south and west Germany which revealed that the characteristics of the risk of these hazards are perceived as similar, e.g. well known and of less personal vulnerability. Thus, the consequences a hazard can have or impose on the social system depends strongly on the personal reaction and preparedness to prevent damages; it is not only a matter of the political and administrative body. "The risk perception is a fundamental base for the decisions and behaviour concerning natural risks and their management of natural risks. Consequently, the risk perception of the inhabitants of a community has been taken into consideration concerning disaster management planning at community level. For the development of effective information strategies on protective measurements (risk and communications policies), the risk perception of the targeted group and as well influences on risk perception should be known" (Plapp 2001, p. 2).

Taking these remarks into account, the results of the investigation conducted by Kaiser (2006) have shown that risk perception is a function of time, but not necessarily correlated with higher awareness resulting in better self-preparedness or precautionary action. People who experienced the disasters of 1953 or 1962 are more aware of the risk living in low-lying areas, but this did not necessarily lead to precautionary measures (Kaiser 2006). A comprehensive study about risk perception and the detailed results for the southern North Sea can be found in Kaiser et al. (2004).

Recently, the multidisciplinary research project "Climate Change and preventive Risk and Coastal Protection Management at the German North Sea coast (KRIM)" was completed (Schuchardt and Schirmer 2007). The project consortium consisting of different disciplines such as economy, social science and engineering has addressed the questions: What are the requirements of coastal protection to be integrated in ICZM under the boundary conditions of an accelerated sea level rise and increased storminess? And what are the interpreting pattern and decision-making processes that influence the integration process? The KRIM project was divided into three parts: (a) political-administrative risk, (b) scientific risk, (c) public risk. Each risk approach has been treated by one or more sub-projects. The main results of the sub project "Climate Change and Public" are congruent with the findings published by Kaiser (2006). Almost 95% of the interviewees agreed that the consequences of climate change will force increased efforts in coastal protection (Peters and Heinrichs 2007, p. 126). "The analysis of cognitive reactions of test persons reading an article about coastal protection confirms the risk perception of storm surges as a relevant risk and a high confidence in present coastal protection" (Peters and Heinrichs 2007, pp. 127–128). One interesting outcome of the investigation was that the risk perception of storm surges is higher in Bremen than in a coastal community bordering the North Sea (Wangerland) and that the confidence in coastal protection is higher in Wangerland than in Bremen. The available data gave no evidence for the reason for these differences in perception between urban and rural areas (Peters and Heinrichs 2007).

3.3.3 Risk Communication

To understand the present situation of risk communication within the political and administrative system the results of the KRIM project will be summarised shortly – see Lange et al. (2005 2007). In Fig. 3.13 the actors in coastal protection are shown.

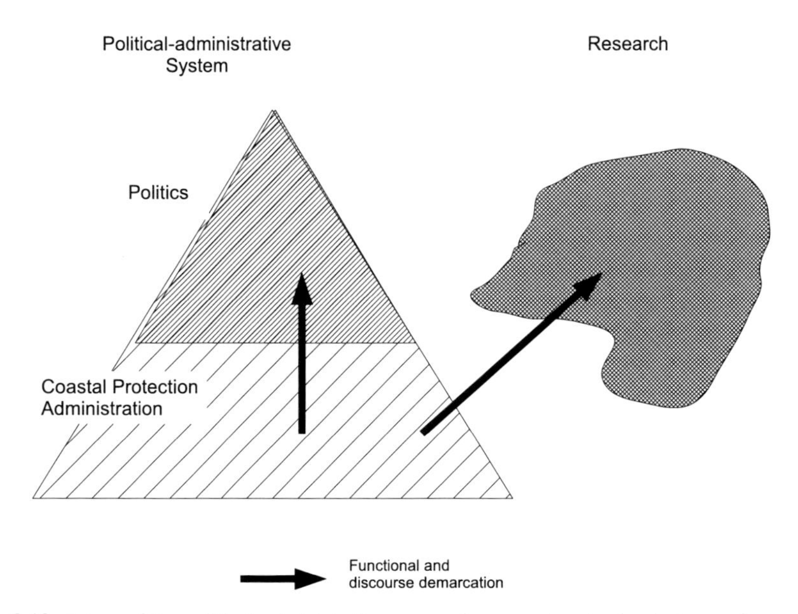

Fig. 3.13 Actors of the political-administrative system for coastal protection in Lower Saxony Source: Lange et al. (2007).

Lange et al. (2007) assign the main task for coastal protection to the coastal protection administration. However, this main actor has certain relationships to politics and scientific research. The differences between administration and research is obvious: administration is embedded in a political decision-making process concerning coastal protection and has to implement decisions. Research aims to understand processes, consequences of impacts and tries to offer reasonable options for action. The administration for coastal protection is located within the area of conflict between the current coastal protection strategy and the results and consequences of climate change research, which are widely known and influence the political as well as the entire social system. These results lead to the question as to whether the present strategies are sufficient for the coming years?

The political system has to decide on reasonable new concepts or strategies. Two knowledge basis are relevant: The accumulated experiences and research results. A new situation results from the knowledge about climate change and the consequences for coastal zones (Lange et al. 2007). The new insights cause potential problems for the political system as well as for the responsible administration, because its own action in the past can be discredited. In many cases the political system will pass the buck onto the administration measures (Lange et al. 2007), which had been taken previously (for example Elbe high fluvial water in 2004). The problems between politics and administration occur because of the differences in the system they are embedded in. For example, administration mainly acts on specific orders from the political system even in cases where the political system should only provide frameworks and objectives. In Lower Saxony the coastal protection administration is divided into two parts: the upper and the lower water and dike administration. The upper administration is the Ministry of Environment and the lower administration is the county. Besides that, the Agency for Water Management, Coastal Protection and Nature Conservation (NLWKN, installed in 2005) acts as the technical authority. The NLWKN is responsible for the design of the embankments, for the calculation of the design water level of the main dikes and for the updating of the Master Plan for Coastal Protection. On the other hand, this technical authority should provide strategical options and alternatives in coastal protection on the mainland as well as for the barrier islands, e.g. with regard to the consequences of climate change. The NLWKN is also the consultant of the dike boards in case of necessary work on their main dike line.

The main result of the investigation of Lange et al. (2007) is, that a discourse about risk only takes place indirectly. The possible additional risks posed by climate change are not acknowledged as “new” risks. The conclusions of the administration are that the risks are manageable, and the required safety can be achieved by scientific-technical optimisation. For this reason the risks will be relativised. According to the scenarios presented by the KRIM project, the reasons to stick to the present strategy are: first, future safety can be guaranteed by optimisation of methods, techniques and strategies, and second that enough time is left to adapt to new circumstances. Questions about new strategies are delegated to the governmental research on coastal protection (e.g. KFKI) and to the political system. New insights into climate change and its possible impacts bear no influence on the existing confidence in the present strategy. Finally, the existing strategy of the coastal protection administration is to delegate the decision on new strategies to the political system on the one hand and to science on the other hand. The expectations are, that these actors provide the basis for a new concept and in the meantime the administration will adhere to the existing strategy (Lange et al. 2007).

The majority of actors in the coastal protection administration prefer to act on the basis of secure knowledge rather than to undertake decisions based on uncertainty. Consequently, action can only be taken if the insight, e.g. into climate change, is empirically sound and acknowledged by the political system. Lange et al. (2007) concludes that this concept of action only on the basis of secure knowledge is highly risky, because assurance can only be gained after an extreme event, i.e. when its too late for preventive measures. Confidence in experience and in the present system is higher than in scientific uncertainty, and this is reflected by the impression of being able to manage and to optimise safety.

Concerning communication about the risks that exist in coastal zones there are many ways of involving stakeholders or even the general public in the process of risk management or other issues (see Sect. 4.2). This begins with providing information on safety or uncertainty, and continue through consultation or more progressively to the comprehensive participation in decision-making processes. However, one main actor is missing in these projects: the dike boards, which belong to both the general public, because everybody living in the floodplain is a member and to the responsible actors. The dike boards play an important role in the political-administrative system because they are responsible for maintenance and improvement of the main dikes, and to some extent they have to raise awareness within their protected area for the problems that exist or would arise in future (see Sect. 3.4). Do the dike boards have a communication strategy and how effective is the strategy? These questions will be published in a diploma thesis by Lampe (2008) for the Lower Saxonian dike boards.

The German Risk Commission concluded in the final report on the review of risk analysis and structures that the following were in place (Risk Commission 2003):

- Clear differentiation between risk assessment (scientific analysis) and risk management (declaration of necessary actions and evaluation of possible measures to reduce risks).
- Common and transparent procedures to assess and to manage risk.
- Appropriate involvement of relevant target groups in the decision-making process.
- Acknowledgment of risk communication as part of risk management.

Thus, the recommendations to involve relevant target groups (stakeholders), or even the broad public is also emphasised in the context of risk management. Besides that, the existing instruments on international and European level (see Sect. 4.2) also emphasis the active involvement of stakeholders within the decision-making processes. The final report of the Risk Commission in Germany identified several problems and gaps within risk communication. As a consequence the project "Development of a multi-stage process of risk communication" was initiated (Renn et al. 2005). Risk communication is the mutual exchange of information and arguments based on the following defined principles (Renn et al. 2005, p. 11):

- Objective clarification of the state-of-the-art in scientific research about the effects and consequences of events, substances and activities on the environment and health.
- Coordination of the actors and the information of the affected public with regard to possible protection measures and adaptation of the behaviour.
- Comprehensive information about the applied procedures to assess and to balance risk.
- Clarification of the position of relevant target groups.
- Provision and application of appropriate communication methods considering the problem-oriented and democratic involvement of different actors in the process of risk assessment.

Furtherone, Renn et al. (2005) divide risk communication into a horizontal and a vertical part. The horizontal part belongs to the assessment, management and evaluation of risk, and the vertical part encompasses the appropriate involvement of decision-makers, stakeholders and even the broad public. Within the report the different aspects of risk communication were exemplified in institutional (or expert) communication and stakeholder communication. The report formulates basic requirements for comprehensive communication (p. 14, see also e.g. Wilson et al. 2003):

- Clear, early and traceable documentation of all assessment processes and results containing information about applied evaluation procedures and criteria as well as the objectives and the legal framework.
- Details on how comments and remarks have been processed.
- Information about the possibilities of participation and co-determination.
- Provisions to allow feedback.
- Information on public events and dialogues concerning the risk in question.
- Information on available literature and statements.

Thus, these basic requirements for risk communication provide a similar framework as the basic principles for progressive stakeholder engagement in ICZM. Consequently, the underlying principle is based on appropriate communication and cooperation in decision-making processes – see also Renn and Webler (1994). The differences between risk communication and the purpose of this dissertation is the wider approach of sustainable coastal development, which comprises various risks and uncertainties (not all of which will be tackled in this dissertation). Focussing on coastal protection zones, risk plays an inherent role. The following section will briefly outline an integrated risk management approach based on several references.

3.3.4 Integrated Risk Management Approach

In the previous section the risk classification based on the WGBU procedure was presented, showing that storm surges belong to the cyclops type (Markau 2003). The WBGU identified three main strategies to address open questions pertaining to this risk type. Another question tackled by the WBGU is the possibility of transferring one risk type (e.g. cyclops) from a certain risk area to another. In Fig. 3.14 is shown that the transfer of risk types within a certain areas as well as from one area to another might be possible. Consequently, the question is: If storm surges are of the cyclops type, what has to be done to change the risk area? The probability of

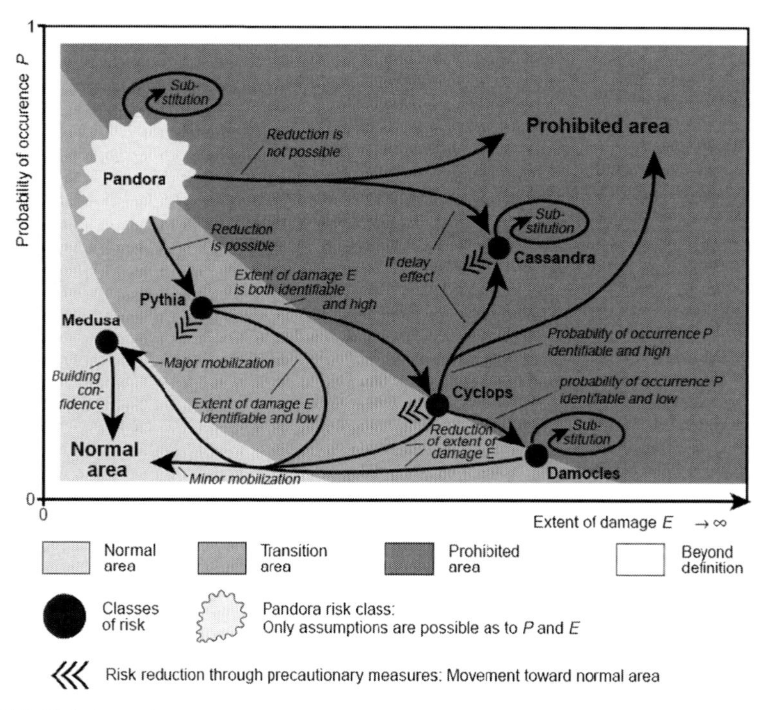

Fig. 3.14 Risk dynamics
Source: WBGU (2000).

occurrence of a storm surge can not be influenced. So, following the diagramme, it is necessary to reduce the extent of damage. This can be achieved by precautionary measures that enable the transfer to the normal risk area. What are the consequences of this approach to coastal protection?

Risk according to coastal protection can be defined as in formula (3.1). For the determination of the "probability of failure" the design water level and the construction of the embankment have to be considered, e.g. Kortenhaus (2003). In case of failure the vulnerability is linked to the extent of the damage. The extent of the damage depends on several items some of which have already been mentioned (see Sect. 3.3.1). The design water level has to be calculated as described in Sect. 2.3.2 and contains uncertainties. The significance of these uncertainties inherent in the determination for decisions on coastal protection are comprehensively described e.g. in Kunz (2004b). Associated with the design water level is the probability of failure for a main dike; this probability should be as low as possible and it should be the same for the entire German coast. The same probability of failure leads to different risks, because there are differences in the vulnerability of rural and urban areas. As already mentioned, in The Netherlands the concept acknowledged the different degree of vulnerability with the different safety levels ranging from 1 to 10,000 (Delta area) to 1 to 1,250 along the rivers.

The design water level in Germany is the result of the societal decision-making process. The duty of coastal protection is the protection of people and their property against flooding. This societal demand implicates, that the embankments should be built as safe as possible. But, the question is how to define the term "as safe as possible"? The question has to be answered with a view to different considerations: the flood-prone area has to be protected against flooding by a technical structure (i.e. main dike) which height is determined by taking into account the highest storm tide, the wave run-up and the fact that the guarantee of absolute safety is not possible. Obviously, the main problem is that the highest storm tide can not be determined and all attempts to do so display inherent uncertainties. A safety standard has to be defined which is able to fulfil the subjective impression of safety for the public and provide an applicable safety standard for a construction. The design water level should fulfil these requirements. Consequently, the safety standard is not defined for the flood-prone area, it is defined for the technical structure. Over the last decades the safety standard, as defined by the Engineering Committee, has met public requirements. The safety standard is incorporated in a protection concept which guarantees a defined safety or accepted risk and reasonable costs for the implementation (Kunz 2004b).

In this context, Giszas (2003) and Kunz (2004b) refer to the ALARP concept (As Low As Reasonably Practicable), which is a modified version of the risk classification by the WBGU. The ALARP concept divides the risk into three main groups: (a) acceptable risk, (b) tolerable risk and (c) non-tolerable risk. The ALARP area is the area between acceptable risk on the one hand and non-tolerable risk on the other hand. Kunz (2004b) divides the technical part of the protection approach into: (a) definition of the decisive high water level, (b) definition of the decisive wave run-up and acceptable overtopping rate, (c) periodical inspection and improvement of safety status, (d) technical standards and recommendations for the construction and maintenance of coastal protection infrastructure and (e) reduction of risk by the introduction of additional embankments. Besides that, non-technical measures also have to be considered, like contingency planning and public information. At the moment, the concept of coastal protection at the German North Sea coast, i.e. the design of a single line, leads to the strengthening and heightening of this line. This is caused by items (a)–(d) of the technical concept. Item (e) has been acknowledged by the State Law on Dikes (§29), but in practice this is not enforced (see Sect. 2.3.2). The expectations of the inhabitants and the politician in the flood-prone are that the agreed safety standard will be guaranteed, which under changing circumstances lead to the necessary adaptation and improvement of the embankments, i.e. mainly heightening of dikes. If this reactive management strategy is not be enhanced by additional measurements in the hinterland, additional problems might occur (Kunz 2004b, p. 268):

- The reactive strategy will cause an increase of vulnerability, because of increasing high water levels and the failure of a main dike can not be eliminated. Further development in the flood-prone area will also increase the vulnerability.

- The concentration on the limited budget caused the heightening of the main dike and the disregard of the second dike lines.
- The strategy to continuously improve the existing line of defence to increasing sea levels will tie up almost all the available budget.
- The current strategy implicates the impression that additional measurements like self-preparedness in case of failure are unnecessary.
- The tight government budget will give rise to the argumentation that no alternative to the current strategy is available.
- According to international agreements the reorientation to integrative management needs to consider the protection of the people, the development and the conservation of values and the principle of sustainability. For coastal protection a shift of paradigms from "defence" (single line of defence) to risk management (spatial protection concepts) will be necessary. The consequence will be that the safety standard has to be adapted to the required sustainable development of a certain area.

Although, the risk approach has not been applied, e.g. to the determination of the design water level yet, due to many uncertainties, these approaches are already the subject of basic scientific research – see e.g. Oumeraci and Kortenhaus (2002), Kortenhaus (2003), Elsner et al. (2004), Mai (2004), Kortenhaus et al. (2007).

In short, an integrated risk management approach is required. Within the KRIM project an integrated risk management approach has been applied, i.e. improving cooperation and collaboration between different agencies responsible for risk reduction in the coastal zone, especially for coastal protection. Markau (2003) suggests a risk concept to deal adequately with natural hazards in coastal regions: (a) scientific risk analysis (investigation of states), (b) socio-political risk evaluation (assessment of social effects) and (c) economic-political risk management (optimisation of risk situation) – see Fig. 3.15. The main point is that these different approaches do overlap in respect of information and applied methods. Cooperation and communication is limited because these fields stick to their traditional scope of responsibilities (Markau 2003). The same has been diagnosed for the status of the political-administrative system responsible for coastal protection in Lower Saxony and Bremen (Lange et al. 2005). The disadvantages of the sectoral approach are the divergent aims and the difficulty in building a consensus on certain tasks in planning. The advantages of the integrative risk approach are the early cooperation on and communication of problems and challenges. Also the reduction of the knowledge deficit and exchange of information to tackle the risk is rated as an advantage (Markau 2003). Problems or obstacles within an integrated approach are the possible interruption or the intentional time delay of the process by an involved institution. Markau (2003) suggests to divide the sectors into analysis, evaluation and management within the integrative approach because each sector has its specialists which provide high quality data and information for the process. This approach has been applied at the KRIM project, and the experience was that the effort needed to generate cooperation between the different sectors should not be underestimated. This experience is also valid for the multidisciplinary project ComCoast where different branches such as administration, engineering companies, universities and author-

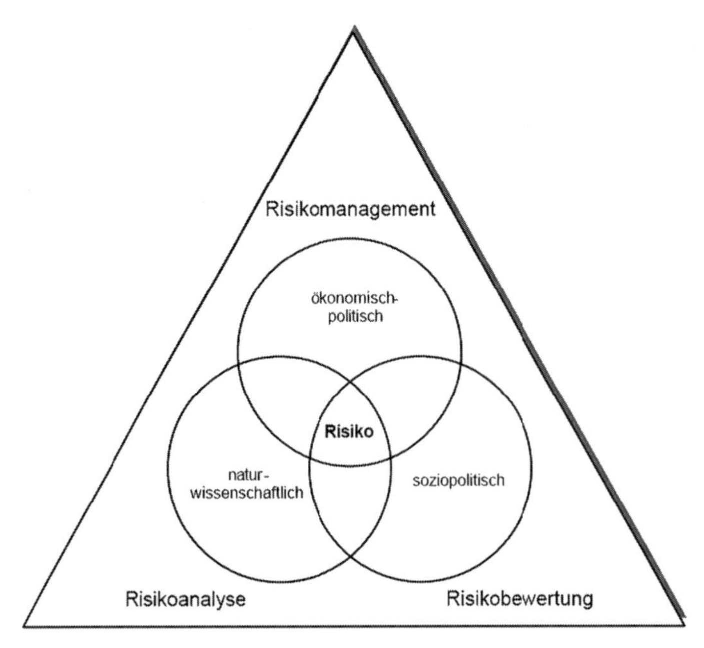

Fig. 3.15 Integrated risk approach
Source: Markau (2003).

ities have worked together in four different thematic sub projects (RWS 2007a). Efficient and effective cooperation and communication between all parts requires a big effort.

A basic insight into the necessity of participation is given in Renn et al. (2005, p. 99): "The larger the number of individuals and groups that have the opportunity to actively participate in risk regulation, the greater is the chance that they develop trust in the institutions of risk regulation and also assume responsibility themselves. However, participation cannot and must not substitute for effective and timely risk management. It should proceed parallel to and along with the prescribed regulation process. Above all, the participation process should not obscure or diminish the responsibility of legal decision makers. Participation within the framework of collectively binding regulations serves to prepare and help in decision-making processes, but not to distribute the responsibility among many (if possible anonymous) shoulders."

To conclude, the advantages and disadvantages of integrated approaches are valid for many projects dealing with cross-sectoral tasks. These experiences have been made and published within risk management, disaster management and also in the investigation of MCPZ. Therefore, within the process of the implementation of ICZM projects the focus has to be on integration and in particular on cooperation and communication. Parallel structures in vulnerability assessment or compilation and assessment of natural hazards have already been explored in Daschkeit (2007).

3.4 New Safety Needs – Demands for Action

3.4.1 Safety Demands in Germany

The Master Plan for Coastal Protection Lower Saxony/Bremen (Mainland) indicates that approx. 125 km of dike line have to be strengthened in Lower Saxony (NLWKN 2007b). In Fig. 3.16a an overview is given of the necessary measures at the coast in Lower Saxony and Bremen. The focus area of coastal protection projects is around the Jade Bay and at the tidal river Weser. The red lines indicate necessary

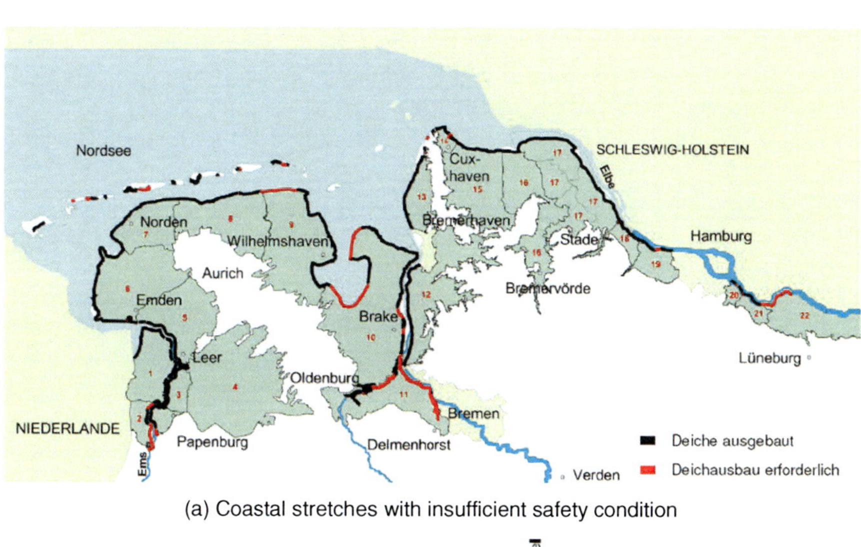

(a) Coastal stretches with insufficient safety condition

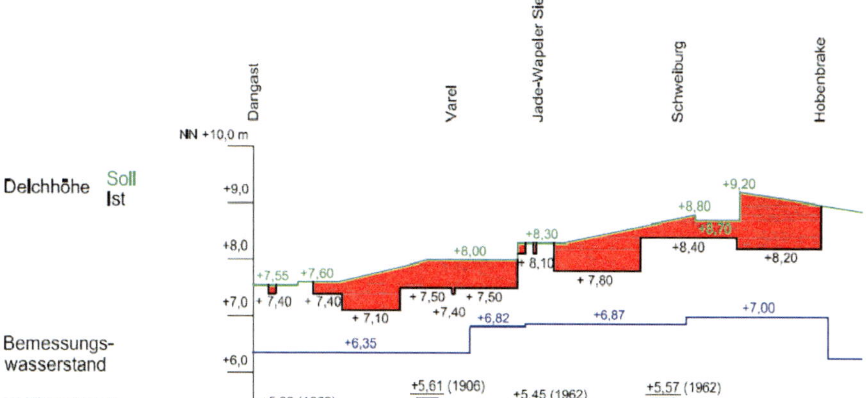

(b) Cross section of the southern part of the Jade Bay

Fig. 3.16 Overview about sites of action in Lower Saxony and Bremen
Source: NLWKN (2007b).

work on coastal protection elements and the black line indicates that coastal protection elements are adequate in state. In Fig. 3.16b a cross section of the southern part of the Jade Bay is shown. The cross section gives a detailed overview on the current situation of the dike and the required height. The figure shows that in some parts the crest height of the dike is almost 1 m lower than required. Furthermore, this coast line has a peaty subsoil, an important road on the inside of the dike and the hinterland is about 1.5 m below the dike foot. For example, a study of the area including land use investigation and the options for coastal protection to improve the dike line in the southern part of the Jade Bay was prepared by Jeschke (2004).

In Fig. 3.17a the present condition of the main dike is shown. In 1962 the height of the main dike line was sufficient, but afterwards this dike line was heightened. Because of the peaty subsoil and the floating peat in front of the dike the only solution was to install sheet piles. Nowadays, this stretch also needs to be strengthened and because of the bad condition of the subsoil 20 m long sheet piles will be installed on the inner slope of the main dike (see Fig. 3.17b). The installation of these sheet piles (about 300 m) took about five weeks; a complicated procedure. On the other hand, the dike was drained for over half a year, because of the moisture in the dike body.

For the western part of the peninsula Butjadingen the dike is indicated by a red line, because the clay layer of the dike is not strong enough. Along the tidal river Weser main parts of the dike crest are too low. Different concepts to improve these sections have been investigated in the research projects KLIMU (Schuchardt and Schirmer 2005b) and KRIM (Schuchardt and Schirmer 2007).

Finally, the main problems with the coastal protection system are situated in highly vulnerable areas, around the Jade Bay with inappropriate soil conditions. Furthermore, parts of the coast line at the eastern part of the Jade Bay has no or a very narrow fore land and the eastern part of the Jade Bay is the main target for westerly storm surges. In East Frisia the coastal protection system has been adapted to the required safety standard over the last decades. The last big coastal protection project in East Frisia, the Ley Bay project, has already been completed (see Box on p. 160/161 and Fig. 5.24).

3.4.2 *Safety Demands in The Netherlands*

In The Netherlands the urgent sites have already been identified as "Weak Spots" ("Zwakke Schakels", MVenW 2002, Provincie Noord-Holland 2005). Weak Spots are stretches of the coast line which have inadequate safety conditions. The priority categories for the Weak Spots are a combination of safety and spatial quality (see Fig. 3.18). To determine the spatial quality of an area, several indicators have been elaborated: e.g. spatial variety, economical and societal functions, cultural functions, sustainability and attractiveness (MVenW 2002). The issue safety was divided into four types: coastal sediment management (kustlijnhandhaving), wave overtopping, safety against overflow (appropriate adaptation of water drainage system) and risk

(a) Sheet piles on the dike crest

(b) Sheet piles on the inner slope

Fig. 3.17 Sheet piles at the eastern part of the Jade Bay
Source: Ahlhorn (1997, 2005) Frank Ahlhorn.

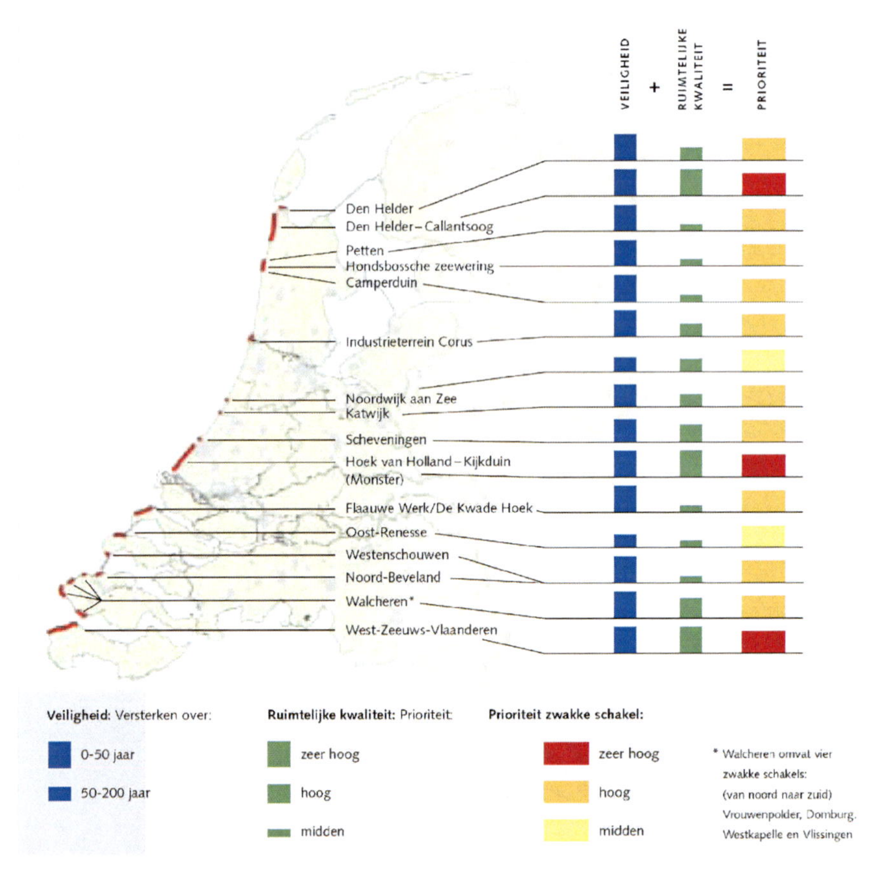

Fig. 3.18 Weak Spots in the Netherlands
Source: MVenW (2002).

management. A detailed programme to improve the Weak Spots and thorough investigations of the dike ring areas has been launched – see e.g. MVenW (2005a). The approach of the FLORIS study (see Sect. 2.4.2) was to develop and apply new methods for testing flood risk to sixteen dike ring areas in The Netherlands. The result of this study was that some dike ring areas failed to meet the required probability of flooding (MVenW 2005a). But, the results of this study are seen as first steps on the way to a probabilistic approach to determine flood risk for dike ring areas in The Netherlands.

The Pettemer and Hondsbossche Sea Defence is one of these Weak Spots. It is located in the province North-Holland. The sea defence consists of a dike between coastal dunes. The current crest height of the dike is between 11 and 13 m above Mean Sea Level (here NAP, Dutch Water Level). The calculated crest height for the year 2060 ranges between 13 and 16 m above MSL (Provincie Noord-Holland 2005 and Smit et al. 2005).

The stimulus to improve or to strengthen the main defence line is based on periodically conducted tests. According to these testings and to the FLORIS project some dike ring areas do not fulfill the safety standard. The design of dikes is for a life time period of 50 years. In case of costly hydraulic constructions a period of 100 or 200 years will be projected. Testing relates to a time horizon of 5 years. The policy in The Netherlands can be summarised as: "designing robustly and testing sharply" (personal communication Frans Hamer, project leader SBW in The Netherlands).

3.4.3 Looking Back and Forward

A consequence of the percolating of environmental issues and the early climate change discussion was the installation of the **I**ntergovernmental Panel on Climate Change (IPCC) by the UN. The first part of this chapter summarised current information about climate change and the likely consequences it can impose on our living environment. The consequences are not limited to sea level rise. Climate change also affects various sectors, like agriculture and tourism. The first assessment report of the IPCC carefully states that climate change may happen (IPCC 1990); the latest assessment report explicitly states that the monitored changes are partly caused by human activity, e.g. Solomon et al. (2007), Parry et al. (2007), Barker et al. (2007), IPCC (2007). Inherent to climate change is the fact that the extent and the consequences are likely, i.e. uncertainty exists about these aspects. Large research programmes deal with the investigation about the extent of climate change and its consequences and the solutions which may be applied to mitigate or adapt to these changes, projects in Germany e.g. Jonas et al. (2005), Spekat et al. (2007), Grossmann et al. (2007), von Storch and Weisse (2008).

Since the installation of the IPCC and the Earth Summit in Rio, the principle of sustainable development has reached the political agenda (see Sect. 1.1). The main questions are how to integrate the three crucial elements (social, economic, ecologic) to achieve sustainable development and how to apply this principle to the coastal zone?

Daschkeit (2004) explained that these questions were transferred from the political field (Earth Summit) to the scientific field. Science asked the question what do we have to know to achieve sustainable development? Growing awareness reveals that the threats on coastal zones increase due to an acceleration of the sea level rise. Consequently, Integrated Coastal Zone Management (ICZM) was introduced within the document about the Earth Summit in Chap. 17. Again, at the beginning of ICZM the question was: what does it mean? Is ICZM a process, a procedure, is ICZM an advanced spatial planning process? A key challenge is how to integrate different sectors in coastal zones. New processes like the Agenda 21 process have been initiated to achieve sustainable development (participation of stakeholders). The initiation of participation processes grew to a backbone of integrated management processes. Additionally, forecasting and estimation of future developments became important tools in trying to find the right strategies, i.e. adaptation and mitigation. The attitude "do nothing" has been weakened by all these new approaches and processes. Although, uncertainty is inherent, people try to prepare themselves against possible impacts. All decisions taken are based on uncertainty.

Nature is risky, natural hazards and their devastating consequences make these threats visible. All technical constructions contain risks. People started to estimate the risk of technical constructions e.g. nuclear power plants. The dikes were constructed on uncertainty from the earliest years, when the height of the dike was estimated by intuition and experiences. A dike is a technical structure and therefore it can fail. The determination of the probability of failure is a topic of recent research projects, e.g. TAW (2000), Oumeraci and Kortenhaus (2002), Kortenhaus (2003), Mai (2004), RIKZ (2006), Kortenhaus et al. (2007), D'Eliso (2007), Richwien and Niemeyer (2007). Also, the probability of occurrence of severe storm surges or even extreme storm surges with a very low probability of occurrence have been investigated – see e.g. Jensen and Mudersbach (2005) and box in Sect. 3.1.3. The technical parameters *design water level* and *design wave run-up* were introduced to determine the necessary height of a dike. These parameters were the result of a societal decision-making process. Consequently, the safety standard given by these parameters is not defined for the flood-prone area, it is defined for a technical structure (Kunz 2004b). Thus, the risk approach of coastal protection has been transferred to technical structures in the coastal zone. Because of the vulnerability of coastal zones, especially against storm surges and the consequences of dike-failure both aspects are the subject of basic research. Risk is determined by the probability of failure and the vulnerability of the endangered area. The vulnerability increases because of continuous economic growth. The probability of failure of the technical structures is still not really known.

The present situation in coastal protection, especially in Lower Saxony, is based on a safety strategy. The position of the responsible authorities and government is: our dikes are safe! But what about the probability of a dike failure? The KRIM project revealed that decisions within coastal protection are taken only on the basis of knowledge supposed to be secure. That means that the uncertainties inherent in the impacts of climate change and sea level rise do not lead to a revision and change of the present strategy. On the other hand, the approach of risk management is nowadays based on an integrative approach such as sustainable development or ICZM. The basic principle is to *act jointly* to develop an area, to install emergency plans and to be clear about aims and the accompanied risks.

In Lower Saxony the Master Plan for Coastal Protection highlights the vulnerable coastal stretches, which are about 125 km of the complete dike line (approx. 600 km). The highest vulnerability is around the Jade Bay. This area is mainly below sea level, and the main dikes protecting the hinterland are built partly on peaty subsoil. The Netherlands identified for their country several Weak Spots, which have been given high priority safety rating. Consequently, safety is not fully achieved and is to some extent difficult to achieve and accompanied by high expenditure. The

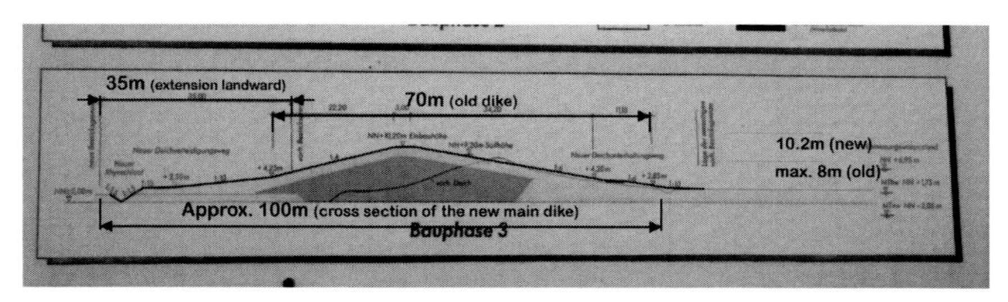

Fig. 3.19 Dike foot extension of the Augustgroden main dike
Source: NLWKN (2007a, b) @ Frank Ahlhorn.

reasons for the necessary improvement of the coastal protection system are the poor condition of one or more elements. The present concept in The Netherlands as well as in Lower Saxony is to heighten and to strengthen the existing main dike line to the required height. The consequences are: increase of the amount of required material, widening of the dike foot, increase of the weight of the dike body and going seaward is not allowed (e.g. National Park) or impossible (main dike without fore land).

The reasons for improvement are almost the same for both countries: increase of storm surges, increase of wave run-up and in some places subsidence of the subsoil. As a consequence, the existing embankments are too low or do not have an adequate cross section. [Remark: The recent heightening of a main dike in the eastern part of the Jade Bay widened the dike foot from approx. 70 m to about 105 m (see Fig 3.19) over almost 10 km.]

The question which poses itself is whether instruments are available to adequately deal with the new circumstances and the effects of the changes?

Chapter 4
Anticipating the Future – Available Tools and Instruments

Contents

This section discusses the existing methods and instruments for consultation and cooperation on coastal protection and the integration within the current statutory framework. The need for new participatory action arises from both the enhanced demand of other types of land use and the need for increased stakeholder involvement within an ICZM process. Consequently, the consideration of different types of land use with specific interests and needs in the coastal zone requires an additional evaluation technique, because different interests and needs have to be weighed and balanced against each other (sustainable development process). By comparison, the traditional responsibility of spatial planning has been limited on balancing and coordinating different types of land use in the coastal zone and to secure several resources, e.g. sand and gravel or areas for recreation.

F. Ahlhorn, *Long-Term Perspective in Coastal Zone Development*,
DOI 10.1007/978-3-642-01774-2_4,

4.1 Legal Processes – Consultation and Cooperation

4.1.1 Relevant Instruments

A detailed and comprehensive description of instruments in nature conservation, water management and spatial planning and the relation to coastal protection can be found in Bosecke (2005). Bosecke investigated the opportunities and challenges of *precautionary* coastal protection considering ICZM, spatial planning, the Habitats Directive, nature conservation and water management. This thorough investigation was prepared for the State of Mecklenburg-Vorpommern adjacent to the Baltic Sea and can not be applied entirely to the situation in Lower Saxony, because of differences in both legislation and the geographical setting.

Basically, the duty of spatial planning is to balance, to secure and to develop land use and the various spatial functions. The mission statement written in §1 Abs. 2 ROG (Federal Spatial Planning Law): "Sustainable development that balances the social and economic interests and needs with the ecological functions and leads to an enduring, balanced and broad based development." A distinction is made between formal and informal spatial planning. Formal spatial planning comprises county and town planning and relevant sectoral plans like water management and traffic. Informal spatial planning comprises framework plans and cooperation agreements. In Fig. 4.1 an overview of the duties and responsibilities of the State of Lower Saxony and the lower administrative bodies is shown.

In 2005, the structure of the government in Lower Saxony was reformed. The former intermediate level between the State government and local administration, the regional government (Bezirksregierungen), was rationalised away. The duties of the regional government were allocated to local administration or to other agencies. In place of the regional government State government agencies (Regierungsvertretung) were established. These agencies are departments of the Ministry for the Interior. They fulfil technical duties of the corresponding ministry. Spatial planning in Lower Saxony is divided into two parts: the upper spatial planning level is allocated to the Ministry of Rural Affairs, Food, Agriculture and Consumer Protection and the lower level to the counties and the cities with county status.

Spatial planning instruments such as the European Spatial Development Perspective (ESDP) on European level include preventive statements (paragraph 142): "Spatial planning at suitable government and administrative levels can play a decisive role here, as well as in the protection of humans and resources against natural disasters. In decisions concerning territorial development, potential risks – such as floods, fires, earthquakes, landslides, erosion, mud flows, avalanches and the expansion of arid zones should be considered. In dealing with risks, it is important, in particular, to take the regional and transnational dimensions into account" (EC 1999). And on national level the ROG §2 paragraph 8 highlights: "[. . .] to care for preventive coastal protection along the coasts and flood defence on the rivers [. . .]". Greiving (2002) describes the duties of spatial planning for risk management as follows: recording of ubiquity of risk (observation), assessment of aggregated risks for a

Administration Level	Instruments/Duties	Spatial Competence
State	*General duties on federal state level:* - Legal Responsibility for Spatial Planning - Landscape Planning and Nature Conservation - Water Management and Coastal Protection	12 nm
State Ministries	*Duties:*	
Ministry for Rural Affairs, Food, Agriculture and Consumer Protection	- Spatial Planning - Administration of Fishery - Agriculture	
State Government Agency	*Duties:*	
Braunschweig, Oldenburg, Hannover/Nienburg, Lüneburg	- Contact point for the public, the regional economy and - Coordination of inter-regional projects and moderation of regional structural development - Improvement of inter-regional cooperation and support to the development of the federal state - Representative of interests of the federal government to the regions and vice versa - Coordination of national and European cooperation - Spatial Planning Concept of the Lower Saxonian Coast (ROKK)	12nm
Special Administration Bodies		
Agency for Geoinformation, State development and Properties	- Fishery - Approval of permission - Coastal Protection	12nm
Agency for Water Management, Coastal Portection and Nature Conservation	- Protection against high water - Coastal protection and island protection - Application of the Water Framework Directive - Dikes and Sluices of Federal State and State - Protection of species and nature conservation (as of 2008 duty of local administration)	
Administration of the National Park Wadden Sea		
County and cities of county status	- lower administration for spatial planning and responsible for regional planning: regional spatial planning programme, responsable for regional planning procedure and planning approval procedure - Planning permission - Landscape framework plan - Approval of land utilisation plan	Mean High Water Level
Muinicipalities	- Land utilisation plan - Town planning - Landscape planning - Dikes	Mean High Water Level

Fig. 4.1 Overview of the responsibilities of administrative bodies in Lower Saxony
Source: adapted from Gee et al. (2006, pp. 78–80).

certain area and preventive assignment of space. Thus, an important method within spatial planning plans and programmes are e.g. the definition and mapping of such specific areas, i.e. priority and precaution areas (for the definition see box below).

Definition – Priority and Precaution Area

Priority Area: An area where a specific duty has priority against others, because of the special situation. In this area all relevant plans and measures have to be considered as priority measure.

Precaution Area: An area where a considerable trans-regional aspect is weighted higher than other competing forms of utilisation.

Important ordinances, plans and laws for the implementation of Multifunctional Coastal Protection Zones (MCPZ) in Lower Saxony are as follows:

- Master Plan for Coastal Protection of 1973, 1997 (for the Regional Government Weser-Ems), 2007 (together with the State of Bremen),
- State Law on Dikes (NDG) 1964, latest adaptation in 2004,
- Ordinance on the responsibilities specified in the State Law on Dikes (ZustVO-Deich 2004),
- Law on Water Management (NWG 2004),
- Law on Nature Conservation (NNatG 2007),
- Law on the Wadden Sea National Park (NWattNPG 2001),
- State Spatial Planning Programme (LROP 1994, latest amendment 2008),
- State Law on Spatial Planning for Lower Saxony (NROG 2007).

The "Ordinance on the responsibilities of the State Law on Dikes" formulates the duties which arise from this State Law. In former times the regional government was responsible e.g. for the determination of the design height of main dikes (Bestick). Now, after the reform of the administration, this is the duty of the NLWKN.

The following section will shortly describe the relevant laws and plans for the Lower Saxonian coast. After this description selected laws, relevant to spatial planning will be investigated concerning their aptitude for the implementation in spatial coastal protection concepts. This comprehensive investigation was conducted from summer 2007 until spring 2008 by Flügel and Dziatzko (2008) within the ComCoast project.

4.1.1.1 State Law on Spatial Planning for Lower Saxony (NROG)

The State Law on Spatial Planning for Lower Saxony is divided into five parts (NROG 2007). The second part sets regulations for the involvement of interested parties within the compilation process of the State and the Regional Spatial Planning Programme (§5). The process foresees a formal participation procedure, i.e. after the announcement of the programme the public has the opportunity to give comments and remarks. But, the draft of the programme is sent other institutions, mainly planning authorities. In §8 the procedure for the compilation of the Regional Spatial Planning Programme is described, therein it is stated that these programmes have to be aligned to regional programmes valid for the specific county and that these programmes have to be developed from the State programme.

4.1.1.2 State Spatial Planning Programme (LROP)

The key principles of the State Spatial Planning Programme (ML 2008) are outlined in article 1.4 "Integrated development of the coast, the islands and the sea". Coastal development should be based on the principles of ICZM, which covered in the first section: sustainable coastal development should be promoted, relevant groups, actors and sectors should be involved in planning and development processes, plans and projects have to be reversible and adaptive to changing circumstances and the control of effectiveness should be carried out for these processes. Through the application of comprehensive balancing and spatial steering approaches land use should be minimised. Article 3 states that the Lower Saxonian coast and the barrier islands have to be protected against damages by storm surges and against erosion. The necessary land area has to be secured. The National Park has to be protected in its uniqueness and its diversity. Touristic infrastructure in the coastal zone has to be secured and to developed. Part 3 of the LROP stipulates the aims and principles of utilisation and the scope for development (Freiraum). This comprises mainly the conservation and development of the nature conservation areas, but also utilisation of these areas by e.g. agriculture, forestry, water management, and coastal protection. Agriculture should be secured and developed as a significant economic user of space in all parts of the State. Comprehensive statements are made on water management and protection against inundation by rivers. In article 10 the following statement is made with regard to coastal protection: "Within the Regional Spatial Planning Programmes areas for dike construction and other coastal protection projects have to be secured in a precautionary way. The projects of coastal protection and high water protection should consider the aspects of settlement development, economy, nature conservation, landscape maintenance, tourism and recreation." After the latest revision, the LROP was extended to the 12 nm zone, for development purposes, i.e. installation of near-shore wind farms.

4.1.1.3 Spatial Planning Concept for the Coast of Lower Saxony (ROKK)

For the first time the Spatial Planning Concept for the Coast of Lower Saxony (Raumordnerisches Konzept Kustenmeer) formulates non-binding statements for spatial planning for the entire coast of Lower Saxony (ML-RVOL 2005). This concept is an amalgamation of available information about all relevant user perspectives in the coastal zone of Lower Saxony. Maps provide an overview of the status-quo, the objectives and the purpose of utilisation. Within the ROKK spatial planning solution approaches are suggested to support and to meet the aims of sustainable coastal development. The special characteristic of the ROKK is that it encompasses the land and the sea side of the coast. It is not the intention of this dissertation to describe or repeat the contents of the ROKK, but some brief remarks should be made on selected aspects (here: tourism and coastal protection), because the ROKK goes beyond the original purpose of spatial planning instruments. The statements concerning several aspects such as tourism, coastal protection, obtaining resources etc. are divided in three parts: description of the status-quo, goals and visions. The basis for the ROKK are the Regional Spatial Planning (amendments not yet finished for all counties) and the State Spatial Planning Programme.

Tourism

The ROKK adopts the principles of the Regional Spatial Planning Programmes relating to tourism, i.e. the priority areas for tourism, especially in the coastal zone and on the barrier islands. Tourism is a significant element of economic development of the Lower Saxonian coast, accounting for approx. 50 Mio. visitors and approx. 1.5 billion € turnover. The ROKK goes beyond the spatial plans in its treatment of the effects tourism can have to other user perspectives and vice versa. For example, it is mentioned that tourism affects itself. The basis for tourism in the coastal zone is an intact natural environment. The State Spatial Planning Programme emphasises sustainable goals in the utilisation of the coastal zone. The goals stipulated by the ROKK are divided into a sea side and a land side part. Specific statements are made for both sides of the coastal zone. The visions section, however, is poorly described, only the "North Sea cycle route" is mentioned as an example of further enhancement of tourism infrastructure, and a "Master Plan North Sea" is touched on briefly. It is acknowledged that off-shore wind farms may serve as touristic attractions, but this has to be elaborated.

Coastal Protection

The first section on coastal protection offers a description of the status-quo. Interlinkages and relationship to other user perspectives are mentioned. To resolve possible conflicts it is proposed that the demand for future coastal protection works be assessed in the context of climate change and an accelerated sea level rise. The recommendations of the project group for the implementation of effective coastal protection management should be applied – see NLÖ (2000). The goals mentioned within the ROKK include: e.g. no further development in inundation areas (natural or defined inundation areas) better contingency plans for the evacuation of the people, via evacuation routes on dams, which are high enough. The ROKK vision is, that the future effects of climate change have to be considered in coastal protection and that future strategies need to account for clay excavation and further threats through a sea level rise by placing restrictions on settlements in the flood-prone area.

4.1.1.4 Law on Nature Conservation Lower Saxony (NNatG)

The aims of the Law on Nature Conservation and Landscape Maintenance (Niedersächsisches Naturschutz Gesetz, NNatG 2007) are given in article 1: Nature and landscape have to be protected, maintained and developed in such a way that the utilisation of natural goods, plants and animals as well as the variety, individuality and beauty of the nature and landscape be protected as a resource for human beings and as a precondition for recreation in the natural environment. These aims refer to the areas within and without settlements. Article 2 recognises that all of these requirements can not be achieved mutually, so the requirements need to be balanced against each other and general interests.

The second part of the NNatG deals with landscape planning as non-legislative technical planning for landscape maintenance and the designated instruments: state landscape programme, landscape framework plan and landscape plan. The third part deals with and provides rules on compensation. In the fourth part soil excavation is regulated (see box). Part five and six deal with protection, maintenance and development of nature and landscape and animals and plants. Here, article 28a is especially relevant for coastal protection zones: in paragraph (1) 4 "dunes, salt marshes and mud flats at the coast and at rivers under tidal influence are highly protected". Paragraph (2) prohibits any deterioration or severe disturbance of these highly protected areas. Exceptions are possible for special reasons and compensation has to be made.

Soil Excavation and the Interlinkage to Coastal Protection
For example, the current concept of heightening the dike requires material, sand and clay. According the Law on the Wadden Sea National Park of Lower Saxony and to most Regional Spatial Planning Programmes the clay should be excavated in the hinterland. The *10 principles for an effective coastal protection*, originally formulated in 1995, reiterates this objective. The amendment of these principles in 2006 lead to a different assessment of the situation (MU 2006), because of the problems of gaining enough clay of sufficient quality in the hinterland, especially for projects planned at the south-east part of the Jade Bay. The original principles were an agreement resulting from conflicts between nature conservation and coastal protection authorities. To solve this problem a project group was installed to develop a new procedure for effective coastal protection in Lower Saxony – see NLÖ (2000). Consequently, the provinces of Friesland and Wesermarsch conducted a survey of suitable areas for the excavation of clay in the hinterland see e.g. LKFriesland (2002).

4.1.1.5 Law on the Wadden Sea National Park in Lower Saxony (NWattNPG)

The National Park Wadden Sea was established in 1986 and the Law on the Wadden Sea National Park came into force in 2001 (NWattNPG 2001). The aim of the National Park is set in §2: "Within the National Park the particularity of the nature and the landscape of the Wadden Sea region should be preserved and protected against interference. The natural processes of these habitats should be maintained. The diversity of animals and plants in the area of the National Park should be preserved." Furthermore, the installation of the National Park should secure the implementation of the EU Bird and the Habitat Directive. Priority habitats in the National Park are e.g. lagoons and coastal dunes; further habitats include e.g. sand plates, the pioneer zone and salt marshes. The Law on the Wadden Sea National Park in Lower Saxony covers the entire National Park, the differentiation in different protection areas, and the forms of utilisation permitted within these areas. The area of the National Park is divided into three zones: restricted zone (highest protection level), intermediate zone and the recreation zone (lowest protection level). Some utilisations are possible in the restricted zone, e.g. grazing and agricultural use, fishery and hunting with defined limitations. Additionally, the area of the entire National

Park has been given the status of a "wetland of international importance" and it has been reported as "important bird area" according to the EU Bird Directive. In 1998 the National Park has been assigned as a FFH area.

4.1.1.6 State Law on Dikes for Lower Saxony (NDG)

The State Law on Dikes for Lower Saxony (NDG) is an unique legislation in the northern part of Germany (MU 2004). The other States, like Schleswig-Holstein or Hamburg, regulate coastal protection via their laws on water management.

The NDG is divided into six parts, but only three of them are relevant for the purpose of this dissertation: part I on main dikes, high water dikes, protection dikes, barriers and dunes, part II on the fore land and safety buffer (Sicherheitsstreifen) and part IV on dike defence, emergency dikes and the second dike line. Coastal protection elements are defined in article 2:

Main dikes are dikes which protect a certain area against storm surges.
High water dikes are dikes which protect a certain area against high water (from rivers).
Protection dikes are dikes situated behind a barrier to protect a certain area against inundation as long as the water can not be discharged due to the closure of the barrier.

An embankment has to be allocated to the above-mentioned dike categories. In article 4 the shape and parts of the dikes are defined. The shape of the dike has to be determined by the responsible authority. The maintenance of the main dikes includes the maintenance of the shape of the dike and of the prevention elements (Schutzwerke) in the fore land or mud flats such as groynes. A stretch of the main dike does not fulfil the necessary safety condition if the crest is 20 cm below the design water level. In article 5 (4) a testing period of 20 years is stipulated, the period is less for highly vulnerable coastal stretches.

Definitions – Safety Elements and Prevention Elements .
For the interpretation of the NDG detailed explanations and comments on specific issues are available in Lüders and Leis (1964):

Safety Elements (Sicherungswerke): Technical measures to ensure the stability of the dike body. The safety elements are part of the dike.
Prevention Elements (Schutzwerke): The intention of prevention elements is to protect the dike body and the safety elements from the physical load of water (currents, waves and ice). Prevention elements in the fore land are to be distinguished from those in the mud flats. The prevention elements are extensions to the dikes.

The experiences of flooding in historical times (from the 1st century on, see Sect. 2.3) are reflected in article 6: "The land owner in the protected area of the dikes and barriers are obliged to jointly maintain the dike (Deichpflicht)." The dike boards are responsible for the improvement and the maintenance of the dikes in Lower Saxony. Today, there are 22 dike boards at the Lower Saxonian coast responsible for an area of approx. 6,500 km^2 and approx. 1.2 million inhabitants. The protected area of a dike board is mapped by certain contour lines in the hinterland determined by the design water level. It is mainly the 5 m contour line, but for some dike boards this can increase up to 8 m (article 9 and annex).

In article 12 the relationship between dike construction and the NWG is described, i.e. whether an EIA is mandatory or not (case-by-case testing). Additionally, organisations are listed which have the opportunity to sue against decisions. The shifting of the dike line may be necessary for a certain reason; in this case the land owner can be compensated. Article 16 formulates that constructions are prohibited within a 50 m buffer on the landward side of the dike. Exceptions are possible. In the autumn and spring of every year visual monitoring of the dikes is to be carried out (article 18).

Part II of the NDG deals with the fore land and the safety buffer. Article 21 (1) states that between the main dike and mean high water level the area with or without a dike has to be kept to a certain width. The land owner and persons with utilisation rights are obliged to maintain the fore land. If erosion of the fore land threatens dike protection then the dike board has to build prevention elements. Article 22 stipulates that the excavation of clay should be done in the fore land. A safety buffer in front of the main dike up to 500 m can be prescribed, if the width is less then 200 m (article 23).

Part IV focusses on the second dike line (article 29): dikes which are able to reduce the inundation of the protected area after a breach of the main dike have to be classed as a second dike as long as they do not fulfil other relevant functions.

Finally, taking the previous descriptions into account, the spatial elements installed in the NDG include:

- The 50 m buffer zone to the landward side of a main dike,
- the 500 m safety buffer in front of a main dike if the fore land is less than 200 m wide,
- the classification of the second dike line to increase the level of protection against inundation.

4.1.1.7 Master Plan for Coastal Protection Lower Saxony – Mainland

The first Master Plan for Coastal Protection in the 1970s (MELF 1973) compiled the strategy for coastal protection for the following years. It comprises a description of the protected area and the projects which have to be carried out to protect the area according the required safety standard. In 1997, the Regional Government of Weser-Ems adapted the existing Master Plan from 1973 for the region between the rivers Ems and Weser to the current circumstances (BR W-E 1997). Recently, the Master Plan for Coastal Protection was published (NLWKN 2007b). This Master Plan covers the entire coast of Lower Saxony and amended the last Master Plan. For the first time, this Master Plan was compiled together with the State of Bremen. It encompasses the aims of coastal protection, a description of elements of coastal protection, the principles for the determination of dike heights and the financial programme for the implementation of necessary work for the next years.

"Coastal protection has precautionary duties. It is indispensable for the safety of the inhabitants and their living and working areas. The protection against flooding and the application of necessary projects have therefore high priority status" (NLWKN 2007b, p. 13). Due to sea level rise the physical load to the coast is increasing and simultaneously the properties in coastal areas are growing in value. Thus, coastal protection has to be adapted to the changing circumstances and is therefore a continuous duty. The primary aim of coastal protection is to provide safety against flooding. The flood-prone area is divided into the responsible areas of the dike boards; this area is protected by dikes. "The aim is to achieve almost the same safety level along the entire coast of Lower Saxony" (NLWKN 2007b, p. 13). Despite calculations based on the highest water level and local wave run-up with an additional safety margin, higher water levels can not be excluded. Thus, "an absolutely safe coastal protection against flooding is not possible" (NLWKN 2007b, p. 13). The storm tide warning service is an important part of coastal protection. The barrier islands in front of the mainland are a "bastion" against storm tides and therefore have to be maintained. Maintenance and preservation of the dike fore land is important, because of the reduction of the physical load of the water (i.e. waves). At some places at the Lower Saxonian coast second dike lines have to be preserved and maintained. "For the application of coastal protection projects more space is needed. The availability of space, especially in areas of settlement, is important. The consideration of these requirements within town and urban plans will be significant. The changing climatic circumstances and their hydrodynamic and morphodynamic consequences require adaptation of coastal protection elements. The spatial need of these coastal protection strategies and the demands on their implementation have to be considered in spatial planning" (NLWKN 2007b, p. 14).

Section 6 of the Master Plan compiles the information on the design of dikes, e.g. necessary heights and cross-sections of a dike.

4.1.1.8 Trilateral Wadden Sea Cooperation

In 1987, the Common Wadden Sea Secretariat (CWSS) was established in the northwestern part of Lower Saxony to support and to enhance the common protection of the trilateral Wadden Sea area. Before that, each Wadden Sea country, The Netherlands, Germany and Denmark, had its own legislation and rules to protect this special nature reserve. Since then many efforts have been undertaken to strengthen the cooperation and coordination of conservation activities. The present protection area in Denmark is about 1,250 km^2 (the Wadden Sea area is about 1,500 km^2), in

Germany the conservation area is 7,360 km^2 (9,050 km^2) and in The Netherlands approx. 2,600 km^2 (3,900 km^2). So, in total the protected area is approx. 11,200 km^2, i.e. approx. 76% of the total Wadden Sea area (Essink et al. 2005, p. 15).

Since 1994, every 5 years a Quality Status Report (QSR) is prepared. The QSR describes the status of the ecosystems within the Wadden Sea and the influences and effects of human utilisation. Recommendations are made for monitoring, assessment and management of the trilateral Wadden Sea area. The QSR 1999 acted as a basis for the Trilateral Wadden Sea Plan (WSP) – CWSS (1998).

The latest QSR was published in 2004 (Essink et al. 2005). Recommendations in the QSR 2004 were made for example for the management of salt marshes (Bakker et al. 2005, pp. 177–178):

- Increased area of natural salt marshes: It is recommended not to disturb the geomorphology of naturally developing marshes as such nor the areas adjacent to sedimentation fields. ... Increase of the area of (semi-)natural salt marshes may take place by breaching summer dikes or sand dikes protecting summer polders. Wherever possible this technique should be applied further. It is under discussion as to whether new salt marshes resulting from de-embankment may include man-made creek-systems and livestock grazing regimes. [...]
- An improved natural vegetation structure of artificial salt marshes including the pioneer zone: It is recommended to specify Target 3 on "natural vegetation structure" of artificial salt marshes as follows: "The aim is a salt marsh vegetation diversity reflecting the geomorphological conditions of the habitat." [...]
- Favorable conditions for birds: Management of salt marshes can be a tool to achieve a favorable conservation status for birds.

Further items of the QSR 2004 encompass e.g. eutrophication, climate and hazardous-substances and several human activities such as coastal protection, shipping and tourism and recreation (Essink et al. 2005).

> The Wadden Sea Plan has been adopted in order to further substantiate joint coherent protection. The principles of sustainable development and use of the Wadden Sea including an important weighting of the relevant interests and avoiding the impairment of traditional interests of the local population are cornerstones in all national, regional and local regulations, policies and management with regard to the protection of the Wadden Sea. The necessity of coastal protection and the safety of the local population is legally implemented in all three countries and has been further specified in national policy and management (CWSS 1998, §1).

The WSP outlines specific targets for the coherent protection and management, for example of landscape and culture, quality of water, sediment and biota and specific area of the Wadden Sea (e.g. salt marshes, tidal area). The target of the WSP for rural areas is to create "favorable conditions for flora and fauna, especially migrating and breeding birds" (Essink et al. 2005, p. 15).

Further European Directives such as the Bird Directive, the Habitat Directive, the Water Framework Directive and the proposed Flood Risk Management Directive and legal instruments in Lower Saxony are compiled in Appendix A.

4.1.2 Potentials and Challenges for Coastal Protection Zones

Within the previous section, selected legal instruments and laws have been described. The options and challenges of coastal protection with respect to a different strategy have already been mentioned in Sects. 3.3 and 3.4. The main point in changing the "single line concept" into a "spatial concept" needs to be adequately integrated within the spatial planning system. Until recently, spatial planning incorporated coastal protection as a specific sector ensuring the safety of people in low-lying areas. The question now is: Do the existing instruments of spatial planning provide the means of implementing spatial coastal protection concepts? What are the gaps and bottle-necks in spatial planning? Has the challenge "climate change" been fully integrated in spatial planning for the coastal zones? To answer this question a comprehensive investigation was conducted elaborating the gaps and opportunities of coastal protection zones in Lower Saxony with regard to the existing instruments in spatial planning – see Flügel and Dziatzko (2008). This investigation was conducted on the basis of the experiences and findings within the German part of the ComCoast project. The following sections summarise the approach and the main findings of this investigation. The results of this study display not only what kind of legal framework is required, but also which methods have to be enhanced to implement spatial coastal protection concepts in Lower Saxony.

4.1.2.1 Investigation Approach

To explore the gaps and bottle-necks within spatial planning instruments, adequate requirements have to be formulated. From these requirements, indicators can be deduced, which will be applied to current instruments. The indicators are measurable units designed to detect a certain state of a system or a project. The indicator approach is widely applied, see e.g. the introduction and application of ICZM indicators to determine the state or progress in ICZM: for Europe see e.g. Olsen (2003), Kristensen (2003), Henocque (2003) and Pickaver et al. (2004); on international level see e.g. UNESCO (2003).

Flügel and Dziatzko (2008) defined the implementation of Multifunctional Coastal Protection Zones (MCPZ) as the main goal. To achieve this goal, interim aims have been defined concerning: (a) the integration of relevant sectoral planning procedures and relevant actors, (b) the integration of land and sea and (c) the consideration of future development and flexibility.

The implementation of MCPZ requires spatial coastal protection concepts such as the approaches displayed within the ComCoast project (see Sect. 5.1.2). Consequently, widening the single line of defence to defence zones demands space behind or in front of the single (main) dike line. Conflicts already exist with regard to the utilisation of the salt marshes (fore land) and have been touched on in Sect. 2.1. Furthermore, the enhancement of coastal protection into the hinterland will also cause problems with existing types of land use, e.g. limited utilisation of marsh land for agriculture. Thus, the implementation of coastal protection zones demands the integration of relevant sectors and actors in the coastal zone to minimise the conflict potential as early as possible. On the other hand, legal instruments are needed to designate a specific area for the implementation of coastal protection zones and these instruments should comprise both land and sea areas. Besides that, with regard to future development and changes, e.g. climate change and its impacts, it is important to provide flexible adaptation options to these changing circumstances as fast as possible.

The indicators proposed for the first interim aim of integration of relevant sectoral planning and relevant actors are "cooperation" and "participation". The indicator *cooperation* indicates the practical extent of cooperation between institutions and whether appropriate instruments are available. The indicator *participation* displays the cooperation between institution/government and the public/private sector, and indicates to what an extent *participation* is incorporated in present instruments.

The "area of validity" was identified as an indicator for the interim aim of "integration of land and sea". Actually, the spatial planning system on land is comprehensive and well-established, but for the sea side the planning system is based more on case-by-case permits or approval procedures than on long-term planning. However, first steps with regard to an introduction of maritime spatial planning has been made, see e.g. Erbguth (2003) and Buchholz (2004). In the previous section it was already mentioned that the LROP has been extended to the 12 nm zone instead of the former border, the mean high water line. So, for the implementation of MCPZ it is crucial to know if integration is already in place and to what extent.

To react adequately to changes imposed e.g. by climate change, but also imposed by other sectors like nature conservation or agriculture, present instruments should provide the opportunity to anticipate future development and to react flexibly to new circumstances. For example, the amendment of the LROP in Lower Saxony was initiated because of the planned installation of a near-shore wind farm. This led to a long process of consultation and information. Hence, the requirement to adapt the present instruments to possible requests in the future and provide the possibility to react quickly on new demands. Therefore, the indicators "monitoring" and "consideration of risk" have been introduced to investigate the state of the interim aim "consideration of future development and flexibility". *Monitoring* has been acknowledged as a necessary aspect to anticipate future development from several user perspectives and to adequately react on changes. The indicator *consideration of risk* refers to the situation already described in Sect. 3.3. Flügel and Dziatzko (2008) investigated how the issue risk was considered in Lower Saxonian planning instruments.

4.1.2.2 Results of the Investigation

The results of the investigations performed by Flügel and Dziatzko (2008) are divided into three parts:

1. SWOT (Strength-Weakness-Options-Threats) analysis of spatial planning instruments,
2. Analysis of selected Regional Spatial Planning Programmes,
3. Structured interviews with selected spatial planning actors and participants of the group attending the discussion process in Nessmersiel.

State Spatial Planning Programme (LROP)

Cooperation – *Strength*: Cooperation is needed to promote integrated development in the coastal zone, projects within large conservation areas have to be aligned to other types of land use; *Weakness*: Cooperation can be widely interpreted.

Participation – *Strength*: In planning and development processes all relevant groups, actors and sectors shall be involved, all aspects concerning coastal zones shall be integrated in planning processes; *Weakness*: Participation is not directly foreseen, decisive proposals on how to involve relevant actors in planning processes are missing.

Area of Validity – *Strength*: Extension of the area of validity to the 12 nm zone, demand for an appropriate development in the proximity of the National Park, the surroundings of the main parts of the biosphere reserve (parts of the National Park) shall be target areas for testing and implementation of sustainable utilisation; *Weakness*: The extension of this programme only regulates the installation of near-shore wind farms.

Monitoring – *Strength*: Plans and projects shall be adaptive and reversible, the control of effectiveness shall support planning and decision processes through comprehensive steering and balancing of spatial needs, land use conflicts shall be minimised as early as possible; *Weakness*: A definition of "control of effectiveness" is missing.

Consideration of Risk – *Strength*: The coast has to be protected against flooding and erosion, the required space has to be secured, research, development and testing of alternative coastal protection strategies shall be considered with respect to climate change; *Weakness*: Measures of protection against high water are more detailed than to coastal protection.

Flügel and Dziatzko (2008) summarised the options for the LROP as follows:

- Cooperation as the basis for further planning steps.
- The aims of the LROP provide a sound basis for the extension of existing structures.
- The demand for participation provides options for the actors to apply tailor-made processes.
- The extension of the area of validity contains potential to minimise land use conflicts in the transition zone.
- Transition zone as joint planning field.
- The principle "plans and projects have to be reversible and adaptive" offers potential of adapting to future developments.
- Legitimacy for coastal protection zones is provided by the statement: "according to climate change the research, development and testing of alternative coastal protection strategies should be considered".

Nevertheless, threats have also been identified: Decision-makers are free to decide on the depth and intensity of the involvement of stakeholders, participation can be limited for a number of reasons (e.g. optimisation of working processes, lack of time, pressure from investors) and land use conflicts are possible due to the restricted focus on near-shore wind farms in the 12 nm area.

State Law on Spatial Planning (NROG)

Cooperation – *Strength*: For the development of the RROP cooperation with municipalities and responsible local authorities for regional planning should be initiated as early as possible. The formal participation process offers the attendance of other actors and the public; *Weakness*: only the highest spatial planning authority has to be involved in the formal participation stage.

Participation – *Strength*: Within the formal participation stage nature conservation organisations, other associations and the broad public have the opportunity to respond, the rules of the NROG provide transparency; *Weakness*: The involvement of the broad public is voluntary.

Monitoring – *Strength*: Significant environmental threats during the compilation of the RROP are monitored. *Weakness*: Only the drafting of the RROP is monitored, there is no monitoring prescribed for the aims of the RROP.

The options for the NROG are as follows:

- Close cooperation between municipalities and other regional planning authorities implies a consideration of existing cooperation structures.
- These cooperation structures can be used to develop a mission statement which is necessary for the implementation of coastal protection zones.
- In the case of the implementation of coastal protection zones a monitoring programme has to be established.

The late involvement of the highest level of spatial planning can lead to new specifications and guidelines not being incorporated or considered only with a time delay. This may lead to problems with the acceptance of these new specifications. The voluntary option of the involvement of the broad public may also lead to problems with acceptance.

Spatial Planning Concept for the Coast of Lower Saxony (ROKK)

Cooperation – *Strength*: Cooperation is the general basis for the development in the coastal zone. For coastal protection it is viewed as horizontal as well as vertical cooperation; *Weakness*: No proposals on how cooperation structures can be created.

Participation – *Strength*: Participation is mentioned as a basic aim. With regard to coastal protection projects the ROKK demands cooperation with other actors beyond merely providing information; *Weakness*: No proposals on how participation structures can be created.

Area of Validity – *Strength*: Significant statements on the territorial sea and spatial planning in the coastal counties are incorporated. This constitutes an unique approach to the coastal zone; *Weakness*: No assignment of space for coastal protection.

Monitoring – *Weakness*: Missing statements on monitoring.

Consideration of Risk – *Strength*: Risk in the light of climate change is acknowledged in the 'visions' section; important remarks on coastal protection zones are included; *Weakness*: Risk is under-estimated and is restricted to the single line of defence.

The options for the ROKK are as follows:

- Flexibility of the ROKK offers timely reaction to new developments.
- Interlinkage between land and sea offers opportunities for the implementation of coastal protection zones.
- Unique approach offers the opportunity to designate interconnected areas.
- The "vision" of the ROKK according to coastal protection refers to the future development and the adaptation of alternative strategies.

The main threats of the ROKK are that it is not legally binding and that there are no clear statements about the involvement of the broad public in planning processes, especially at the coast.

Detailed Investigation About Selected RROP

The RROP of the county of Friesland offers good opportunities for the implementation of spatial coastal protection concepts, because the RROP emphasises e.g. the importance of the second dike line. Good experiences with cooperation between different institutions have already been gained in the county, but the experiences with participation are sparse. However, the county is interested in the introduction and application of participation methods. The interlinkages between sea and land is seen as the duty of the State, and therefore further coordination on higher level is needed. Approaches to the consideration of future development and flexibility already exist via a GIS application. The GIS allows spatial monitoring, but the consideration of changes is difficult because of lengthy planning procedures.

The RROP of the county of Wittmund does not refer to coastal protection and the representative of the responsible spatial planning authority has no doubt about the existing strategy of coastal protection. On the other hand, in this county participation was conducted voluntarily on the drafting of the RROP. Experience already exists with the involvement of the public, but the interest of the public was minor. Cooperation seems to be executed only on authority level. It seems that the challenges of climate change and the background of the ComCoast project dealing with spatial coastal protection concepts is unknown, and therefore no further action seems to be required. With regard to future development and flexibility the same statements are valid as for the county of Friesland.

The RROP of the county Wesermarsch acknowledges that climate change can impose further threats on the present strategy for coastal protection, but no alternative approaches are seen as capable of solving future problems. Cooperation between different institutions already exists, but there is less experience with public participation. The interlinkage between land and sea is seen as duty of the State, and therefore no further remarks are made in the RROP. The potential for the implementation of spatial coastal protection concepts is rated as small, because of the topography of the county, with areas of −2 m below sea level.

To conclude, the remarks of the current RROPs with regard to coastal protection are mainly focussed on the "hold the line strategy". The amendment of the LROP stimulates the amendment of existing RROPs and thus offers the opportunity to incorporate elements of spatial coastal protection concepts within the RROP. The basis for these amendments is given in the LROP (paragraph 1.4 number 12): "Vor dem Hintergrund zu erwartender Klimaveränderungen soll der Erforschung, Entwicklung und Erprobung alternativer Küstenschutzstrategien Rechnung getragen werden (According to climate change the research, developement and testing of alternative coastal protection concepts should be considered)." To secure necessary space for alternative or spatial coastal protection concepts, the required space should be displayed in the plans and programmes as priority or precaution areas. At first, this will only be applicable on the land side. The experiences of cooperation and participation varies between the different regions, but positive approaches are available and these may serve as a basis enhancing existing structures, and providing best-practise projects for others. The application of the indicator *monitoring* shows that there is willingness to monitor and to adapt the existing plans, but the procedures are seen as too complicated and too time-consuming. The indicator *consideration of risk* leads to a differentiated impression within the counties concerned. The likely impacts of climate change have been acknowledged, but the fact that these might have direct effects on coastal protection is less recognised.

4.2 Participatory Action – Involving the Coastal Society

Remark: Parts of the following text have already been published by Ahlhorn and Klenke (2006a). Those parts are indicated by quotation marks in the text.

4.2.1 Participation Approaches – Retrospective

"The history of public participation in decision-making processes can be traced back until the end of the 1960s. The International Covenant on Civil and Political Rights of the United Nations General Assembly in 1966 can be seen as starting point, where Article 19 deals with the 'freedom to seek, receive, and impart information'. From then on many conferences were held and recommendations made on imparting information, consultation and involvement of the public at different statutory levels.

The resolution No. 171 on regions, environment, and participation of the Council of Europe in 1986 can be seen as the main breakthrough at European level, which was adopted at the Standing Conference of Local and Regional Authorities of Europe. Afterwards, in 1989, the European Charter on Environment and Health was adopted and recognised public participation as an important part in the context of environment and health. In 1990, the Directive 90/313/EEC on freedom of access to information on the environment came into force. Parallel to the European development, public participation on an international level was given a new dimension of importance by the introduction of the concept of sustainable development and the Agenda 21 process. The Declaration of the first Earth Summit 1992 in Rio could be seen as a basis for the Århus Convention on public participation in 1998.

The Convention on Access to Information, Public Participation in Decision-making and Access to Justice in Environmental Matters (Århus Convention) constitutes a new kind of environmental agreement: 'It links environmental rights with human rights' (see UN/ECE 1998, p. 1). The Århus Convention has three pillars: access to information, public participation, and access to justice. The first pillar, access to information, is necessary to provide comprehensive, accurate, and up-to-date information in decision-making processes. In Germany, the Law on Environmental Information is intended to fulfil the required actions of this pillar. The second pillar, public participation in decision-making, is divided into three parts: participation in decision-making processes on the part of the public affected or otherwise interested in specific activities (Article 6), participation in development of plans, programmes and policies (Article 7), and participation in the preparation of laws, rules and legally binding norms (Article 9). In Germany, some of these requirements are covered by existing laws and rules, e.g. the statutory right of involvement of the public in a Planning Approval Procedure or within an Environmental Impact Assessment (EIA). Finally, the third pillar, access to justice, should ensure that public participation happens in reality and not only on paper (UN/ECE 1998). In this context, the Participation Directive was introduced in 2003 as a contribution to the implementation of the third pillar of the Århus Convention (EC 2003b).

The Water Framework Directive (WFD) is another part in the framework for public participation on environmental matters (EC 2000). Within the WFD, stakeholders should be engaged, the public should be involved in hearings and the background information should be accessible to all. There are no rules given by the WFD for public participation, but guidelines are provided for different themes, especially for public participation (e.g. EC 2003a). Nevertheless, the WFD is another legally binding rule which strives to involve and engage the public in decision-making processes in Europe.

As an evaluation of the demonstration programmes of Integrated Coastal Zone Management (ICZM), the recommendations of the EU Parliament and Council for the implementation of ICZM in Europe were compiled. The recommendations state that involvement and engagement of the public is one main principle within the concept of ICZM (EC 2002). It was recommended that each member state should prepare a status-quo report about the national strategy on ICZM. The summary of these international and European documents show that there are many approaches towards stimulating and enrolling public participation. As stated in the Århus Convention 'there is no set formula for public participation, but as a minimum it requires effective notice, adequate information, proper procedures, and appropriate taking account of the outcome of public participation. The level of involvement of the public in a particular process depends on a number of factors, including the expected outcome, its scope, who and how many will be affected, whether the result settles matters on a national, regional or local level, and so on' (see UN/ECE 1998, p. 85). [...]

The relevant law for the implementation of coastal protection schemes is the Lower Saxony State Law on Dikes (NDG 1963). Within the original State Law on Dikes from 1963, there is no article which considers public involvement. Because of the high priority of coastal protection after the severe storm surges in 1953 and 1962 there was no necessity to explicitly regulate public involvement. With other upcoming interests and needs, e.g. nature and environmental conservation the requirements to coastal protection have changed. With the implementation of the Law on EIA, the public will be involved in particular projects concerning coastal protection schemes. There are a few examples for involving stakeholders and using participation for resolving conflicts, e.g. Kaul and Reins (2000), NLWK Norden (2003), Striegnitz (2006). A short overview of other examples of public participation related to coastal protection schemes is given in Ahlhorn (2005).

It is clear that the status of participation in implementing coastal protection or flood risk management differs between the partner countries in the ComCoast project. In the UK, the Environment Agency uses the method of public participation to involve and to implement proposed projects. Over the last years, they have prepared a stepwise approach which has been improved in various projects (EA 2005a). In Germany, participatory action is mainly used to resolve urgent conflicts between the different interests of coastal protection and nature conservation in the coastal zone (e.g. Striegnitz 2006). In the pilot area in Lower Saxony, a new approach to public participation in coastal protection schemes was applied. In that case public participation, involving stakeholders, was used to develop options and development solutions for the period till 2050. The applied method was created in the 1970s to deal with environmental problems and to bridge gaps in knowledge and perspectives of different actors (EEA 2001a)" (Ahlhorn and Klenke 2006a, pp. 110–111).

For example, a comprehensive conflict resolution process was proposed by Ahlhorn (1997) concerning dune management on the barrier islands of Texel (The Netherlands) and Norderney (Germany). Before the solution was elaborated a detailed investigation has been conducted and a detailed description of the situation and the boundary conditions were made. Some of the suggested steps have been applied in the German case study Nessmersiel (Meyerdirks and Ahlhorn 2007b).

The public is now much more aware of their rights and the effects decisions have on their lives. Changes in legislation also mean that we have to involve people more in decision-making processes.

Traditionally, most public organisations have made decisions, let people know what they plan to do and then had to defend their decisions against those who did not like them. Most partner countries have followed this, "Decide, Announce, Defend" approach in the past, but now this procedure has become less acceptable.

For example, the experiences of the Environment Agency are summarised as follows (EA 2005a, p. 2):

[...] Following this approach, there is the risk of:

- Relationships and trust breaking down – often involving local politics and the media-making our work more difficult in the future.
- Making decisions without fully understanding relevant issues and reactions, which means that they may not be the most appropriate.
- Interest groups throwing out our preferred decisions, and us having to go back to the drawing board, often at great expense.

Nevertheless, regarding the efforts which have been made to promote public participation on international, EU and national level, there are still many barriers facing participation.

4.2.2 Barriers to Participation

4.2.2.1 The Approach of ComCoast – Guidance

Work Package four of the ComCoast project has had three main objectives: to deliver an overview of participation methods in partner countries (Colbourne 2005), to develop a communication strategy for participation processes for the pilot projects (Stroobandt et al. 2007) and recommendations for public participation based on the experiences and lessons learned of the pilot projects. Much has been done to improve the tools and to engage the public, such as visual techniques to help to explain complex issues, specialist staff and training to help to communicate. And also techniques on how to improve the process of involvement. For example, the "Building Trust with Communities Toolkit" (EA 2005a) which is a collaborative piece of work produced by the Environment Agency in the UK and ComCoast. While working on recommendations on public participation it became clear that within the literature on participation and mediation, a lot is available about good planning and running of processes, but very little information is available about "barriers" to participation. The barriers have been collated through national workshops and an international workshop held in Middelburg in The Netherlands 2006. The workshop discussions and outputs indicate that most barriers are similar in all countries involved. A guide has been put together to help practitioners to overcome barriers to participation – see Houtekamer et al. (2007). The aims of this guide are as follows:

- Provide a background of knowledge to help practitioners through otherwise unsolvable participation problems.
- Provide solutions to each barrier identified to avoid a generic and prescriptive approach.
- Provide case study examples to help visualise how this could be applied in situ.
- Provide a tool that will help to influence.

Lessons learned in the ComCoast project are: (a) It is better to start working with communities early in the decision-making process. The later you leave it, the more likely it is that trust will be broken, (b) the more complex and controversial the work is, the closer the work with the communities concerned has to be, (c) it is vital to plan how to structure the work with the communities, i.e. carefully looking at the reasons for cooperation, which individuals need to be introduced, agreeing on the best way of working together to meet their needs and the needs of the leading organisation and (d) making sure that the team has the right skills and knowledge to confidently engage the public.

Changes in legislation alone will not ensure that there is an improvement in structuring participation. Both the public and the leading organisation that delivers sustainable coastal schemes must believe in the participation process. This requires knowledge and experience of the people delivering the schemes and strategies on the coast. If the barriers to participation can be identified then working on communicating solutions can be started. For example, it has been recognised across a wide range of specialists that a well planned, thoroughly resourced process will take less time, money and resources in the long run, thus helping to deliver schemes quicker and more economically whilst building strong relationships with partners and the community.

4.2.2.2 Barriers – Approach and Description

The approach of Work Package four is to learn from mistakes in the past as well as from good practice and to use these experiences to provide solutions. It was found that barriers can be divided into internal and external, e.g. *Time* is a barrier with different meanings for both within an organisation and between organisations. Nine barriers have been identified and two of them will be explained in more detail, because they occurred in the Participatory Integrated Assessment (PIA) process of the German case study. The nine identified barriers to participation are briefly explained (for a more detailed description see Houtekamer et al.2007):

Time: Within an organisation people think that participation takes more time to finish the project or process. From a general perspective, organisations or institutions which are relevant stakeholders do not have the time to participate or it is not clear when the people should be informed about what.

Staff: "Organisations may lack the necessary skills of organising participatory processes and internal procedures may be too rigid and time-consuming to cope with the dynamic nature of participation processes. Skills of authorities and stakeholders may not be sufficient to lead to a successful participation process automatically" (p. 11).

Money: Within an organisation this barrier is associated with lack of staff and that money for a participation process had not been included in the budget. Generally, spending money on participation processes is perceived by stakeholders as a waste.

Politics: Within an organisation the lack of trust in and support from higher levels. Lack of political commitment from authorities. Short term politics may influence projects, i.e. community and council elections.

Power: The government does not want to relinquish power to the public by public participation. And people (the public) do not believe that they have power. Between organisations there is the fear that power can be shifted within a participation process.

Troublemakers: Internal troublemakers can be obstructing and criticise the process by doing nothing constructive to overcome the critics. The external troublemakers can be politicians, media, interests groups and other involved parties which only obstruct the process without well-founded reasons.

Misunderstanding: An internal misunderstanding arises if no communication and information is provided, unspoken grievance, lack of commitment or a refusal to cooperate are the results. Generally, misunderstanding can have different aspects: too little communication, assuming to know the opinion of others, trust, transparency and crucial misunderstanding of the key roles of the people attending the process (not having the the decision-makers on board).

Bad experience: If there was a bad example in the past, the willingness to run a participation process will be hampered within an organisation. Further, internal problems can be that staff felt not backed or supported by their own organisation. Bad experiences of external attendees include unkept promises.

Closed minds: Within an organisation the experience is that a process has worked well without participation in the past and new methods seem to be dangerous. Outside the organisation conservative ideas or solutions may prevail because of fear of innovations. The main reason for withdrawing the willingness to participate in a process can be the NIMBY-principle (Not In My BackYard). For example, this has been investigated in the ComCoast project with regard to a Dutch case study, where the NIMBY phenomenon combined with the belief that the people do not have really a say in the process appeared (see Roose 2006).

4.2.2.3 Barriers – Two Examples

The following two examples show both how the barriers document is structured and which barriers occured in the German case study. Here, it is exemplified for the barriers *Misunderstanding* and *Closed Minds*.

Misunderstanding

Causes: The causes for *Misunderstanding* can be as follows:

- Not knowing the other person or organisation,
- not speaking the same "language",
- not knowing each others codes, not knowing the rules of conduct,
- not knowing how to interprete the others behaviour,
- failing to realise that *Misunderstanding* may occur,
- not listening.

Solutions: Include a thorough preparation phase and find out more about stakeholders. First, try to use personal communication. One-to-one meetings can deepen the relationship with the person you are talking to and can better be used to express and explain the information at stake. Second, try to develop local knowledge for yourself to help create credibility and win the trust of local people. Making the effort of becoming acquainted with the area concerned can lead to both gaining local knowledge and involving the local public in sharing local knowledge. Third, a clear and precisely explained issue might avoid misunderstanding, hence the need to use appropriate visualisation tools to present your concern.

Tips: Organise site-visits: Specialists get the opportunity to explain the project and get immediate feedback. Local inhabitants can share their knowledge of the site. If starting a project with a cooperation of more than two parties, make a clear and precise project structure, schedule and make clear the rights and responsibilities of each party involved. If you organise an information event, be sure that the facilitator has the right skills and is accepted by the audience as reliable.

Experiences: In Germany, the kick-off meeting started with a comprehensive introduction of the ComCoast project and the aims of the participation process in Nessmersiel. Nevertheless, after the first workshop some attendees had uneasiness about the contents and the intention of the process and the project. They did not express this uneasiness clearly to the leading organisation, it was not directly communicated. In this case the situation was resolved by one-to-one meetings between the leading organisation and these attendees. After these meetings, where the intention and the purpose of the participation process in the pilot area as well as the overall ComCoast project were comprehensively and openly discussed, the attendees agreed to be part of the process again. With hindsight, one mistake has been made, there was no intensive and personal contact with the attendees before the process started. Using the opportunity of personal contact and one-to-one meetings, the expressed concerns can be gathered and treated in planning the participation process.

Closed Minds

Causes: The causes for *Closed Minds* can be as follows:

- Organisations fear open dialogue,
- people's belief that they don't have a real say,
- lack of long-term vision.

Solutions: First, if there are examples of similar problems which have been solved (in an innovative way) then use these examples as case studies to show the possible consequences and the benefit. Second, if you are making suggestions be clear and precise about the methods and instruments and the technique which could be applied to solve the problem. Third, make sure that there is a person with local knowledge and with innovative but not bizarre ideas and concepts to stimulate the attendees. Other solutions may be:

- Try to find people with enthusiasm who are trusted by the organisation to promote your project,
- listen, do not assume,
- show an open mind yourself.

Tips: In practise, try to arrange the process in such a way that two or more solution approaches are possible. One of them could be the conservative one and the others the more innovative ones. If there are *Closed Minds* within the organisation which are against participation then invite them, if possible, to run the process. Further recommendations are: use creative techniques and visualisation tools/diagrams/maps to open minds and really listen to people's concerns and if possible try and use their own quotes and ideas.

Experiences: Within the participation process in the German pilot area the second workshop was intended to deliver patterns for future land use for the year 2050. The circumstances were presented in three scenarios, only the boundary conditions for climate change and the attitude of coastal protection were the same for all scenarios. The attendees were split up in two working groups for the exercise. One working group developed traditional solutions for 2050. The second working group developed more innovative solutions with new ideas. Within the first working group there was a dispute about details of present-day situation and the second group focused more on the year 2050 and the opportunities given by the scenarios. Within the second group there were attendees with local knowledge but also innovative cross-sectoral insights which contributed to the results.

To decide which is the appropriate technique to deliver a participation process for the development of options within a scenario driven process a survey of several participation techniques was made and afterwards an evaluation was done. The next section describes the procedure and the outcome of the evaluation of different participation techniques.

4.2.2.4 Evaluation of Participation Techniques

"Several requirements have to be considered evaluating participation methods. The most important one is to define the aim of the process clearly. Given the aim of the participation process one can determine and derive an adequate technique, e.g. techniques can be used to gather information, other techniques are better for consultation processes, or there are techniques which can be used to work out solutions for urgent problems. Bearing in mind the different qualities of participation processes, i.e. information, consultation or involvement of the public, different techniques are available to meet these requirements. Therefore, the commonly accepted principles for participation has been used as guide to select and to evaluate participation techniques (Ridder and Pahl-Wostl 2005):

- The role of stakeholders should be clearly defined and communicated.
- Stakeholders involved should have visible direct benefits.

- The process should be transparent.
- Stakeholders involved should be representative.
- Stakeholders should be involved from the beginning of the process.
- Stakeholders should receive an adequate and timely feedback showing the results and how their inputs were used.
- Participation should lead to learning and capacity enhancement.

Because of the specific situation in Lower Saxony concerning coastal protection there were additional requirements to be made on a participation technique. The existing embankments have prevented flooding over the last 40 years, even when there were higher storm surges than in 1962. Therefore, the inhabitants enjoy a legitimate feeling of safety behind these embankments and the present strategy. Consequently, thinking about and dealing with new concepts and strategies even under changing circumstances like climate change and an accelerated sea level rise has to be done very careful. A technique was needed to deal with the changing circumstances and the consequences in an open way. The technique should provide the ability to develop a common view of future needs and circumstances and appropriate reactions. Thus, there was a need for a technique that generates a confidential atmosphere between partners" (Ahlhorn and Klenke 2006a, pp. 111–112).

Additionally, taking into account the current assessment of no action required, there was a need to use for scenarios to explore alternative options for the development in the pilot area in future and the stakeholders to provide feedback on the interim and the final results of the socio-economic and ecological valuation method (see Sect. 4.3, Ahlhorn and Klenke 2006a).

With respect to the requirements of interaction and creativity the selection was narrowed to the techniques mentioned below. All considered techniques can be grouped as Participatory Integrated Assessment (PIA) techniques. Comprehensive overviews of participation techniques are given in e.g. EEA (2001a), OECD (2001), Slocum (2003), Cox (2005), Ridder and Pahl-Wostl (2005).

The following techniques were considered to meet the requirements of a tailor-made participation process:

Workshops (WS): Meetings for a limited number of participants to provide detailed information, to discuss and solve problems.

Focus Groups Techniques (FGT): Meeting of invited participants designed to gauge the response to proposed actions and gain a detailed understanding of people's perspective, values and concerns.

Planning Cell (PC): The Planning Cell method engages a restricted number of randomly selected people who work as public consultants for a limited period of time in order to present solutions for a given planning problem.

Policy Exercise Approach (PEA): Meeting of invited participants to synthesise and assess knowledge from various sources and ideas. This approach is scenario-driven in order to assess different alternatives considering the challenges of the scenarios.

4.3 Assessing the Future – Scenarios and Evaluation

4.3.1 Scenarios

The application of the scenario-technique to anticipate future development and especially to create optional adaptation or mitigation strategies is widely used, e.g. IPCC (1990), Pahl-Wostl et al. (1998), Parry (2000), IPCC (2000), EEA (2001b), UNEP-RIVM (2003), Schuchardt and Schirmer (2005b), Wolf and Appel-Kummer (2005), Schuchardt and Schirmer (2007). Scenarios are neither true nor false: They are descriptions of a plausible future development and serve as a spectrum of possibilities and a discussion platform.

The estimation or forecasting of climate change parameters is only possible applying scenarios, which are descriptions of the expected future development. In EEA (2001b) different types of scenarios are described, explaining their purpose and suitability in environmental assessment projects. It was differentiated between qualitative and quantitative and between exploratory and anticipatory scenarios. Qualitative scenarios describe the future development by storylines, preferring visual symbols to figure. In contrast, quantitative scenarios use mainly figures and diagrams to visualise a future development. These scenarios are based on numerical information which has inherent uncertainties and therefore cannot predict the future reliably. Another classification distinguish between exploratory and anticipatory scenarios. Exploratory scenarios start off from the current status-quo and try to explore tendencies for the future. Anticipatory scenarios try to anticipate the future development, and thus the state a certain system might have in the future. So, exploratory scenarios are more or less value-free, because they do not suggest a desirable state in the future (EEA 2001b). The IPCC reports and assessments are based on the Special Report on Emissions Scenarios (SRES, IPCC 2000). Detailed explanation of these SRES can be found in the box below. "Scenarios are images of the future, or alternative futures. They are neither predictions nor forecasts. Rather, each scenario is one alternative image of how the future might unfold. As such they enhance [the] understanding of how systems behave, evolve and interact. They are useful tools for scientific assessments, learning about complex systems behavior and for policymaking and assist in climate change analysis, including climate modeling and the assessment of impacts, adaptation and mitigation" (IPCC 2000). The SRES are described in storylines and encompass different ranges for certain aspects: first, the driving forces like population projections, economic development and structural and technological changes; second, the projections of greenhouse gases and sulfur emissions. The emission scenarios of the SRES are briefly described below (after IPCC 2000 and Solomon et al. 2007, p. 18). The "A" storylines are more economically oriented and the "B" storylines are more environmentally oriented; besides that, the "A1" and "B1" storylines are more global and "A2" and "B2" storylines are more regionally oriented.

The scenario family and storylines described in the box differ from the earlier scenarios used by the IPCC from the First Assessment Report (see IPCC 1990), because new insights and developments in forecasting made an adaptation of the existing scenarios necessary. For example, new findings related to population projections and new insights in correlation between certain aspects like per capita growth and life expectancy (Girod 2006). Also the types of scenarios changed: The first scenarios could be classified as "decision support" scenarios and the SRES scenarios can be classified as "exploratory and decision support scenarios". The difference between these two groups of scenarios is the intention that exploratory scenarios represent different possible futures and decision support scenarios represent more strategic options. Consequently, the SRES scenarios can be classified as explorative approaches with a decision support outcome (EEA 2001b and Girod 2006).

Emission Scenarios of the IPCC: Special Report on Emissions Scenarios (SRES)

A1: The A1 storyline and scenario family describes a future world of rapid economic growth, global population with the highest peak in mid-century and decreases afterwards, and the rapid introduction of new and more efficient technologies. Major underlying themes are convergence among regions, capacity building and increased cultural and social interactions, with a substantial reduction in regional differences in per capita income. The A1 scenario family is divided into three groups which describe alternative development of technological change in the energy system: fossil-intensive (A1FI), non-fossil energy sources (A1T) or a balance across all sources (A1B).

A2: The A2 storyline and scenario family describes a very heterogeneous world. The underlying theme is self-reliance and preservation of local identities. Fertility patterns across regions converge very slowly, which leads to continuously increasing population. Economic development is primarily regionally oriented and per capita economic growth and technological change more fragmented and slower than other storylines.

B1: The B1 storyline and scenario family describes a convergent world with the same global population, with the highest peak in mid-century and decreases afterwards, but with rapid change in economic structures towards a service and information economy, with reductions in material intensity and the introduction of clean and resource-efficient technologies. The emphasis is on global solutions to economic, social and environmental sustainability, including improved equity, but without additional climate initiatives.

B2: The B2 storyline and scenarios family describes a world in which the emphasis is on global solutions to economic, social and environmental sustainability. It is a world with continuously increasing global population, at a rate lower than A2, intermediate levels of economic development, and less rapid and more diverse technological change than in

B1 and A1 storylines. While the scenario is also oriented towards environmental protection and social equity, it focuses on local and regional levels.

The box below describes for example the climate change scenarios for the United Kingdom conducted within the climate change programme established in 1997 (Hulme et al. 2002).

Climate Change Scenarios for the UK

The United Kingdom established a Climate Impacts Research Programme (UKCIP) in 1997 to provide a framework for an integrated national assessment of climate change impacts (Hulme et al. 2002). The UK scenarios for climate change are based on the SRES scenarios A2, A1FI, B1 and B2. The results of the Hadley Centre global climate model are the input parameters for the higher resolution model of the UK. The results of these regional models are the basis for the UKCIP02 emissions scenarios. These emissions scenarios range from *low emissions* to *high emissions*, and are linked to temperature increase between 2.0 up to 3.9°C and a concentration of CO_2 ranging from 525 ppm up to 810 ppm. The consequences for the UK climate and weather are shortly explained: The temperature for the low emissions scenario may rise about 2°C and for the high emissions scenario about 3.5°C. The precipitation may increase between 10 and 20% for the low and for the high emissions scenario up to 35%. It is assumed that the year-to-year variability will be greater in 2020. "Very dry summers such as in 1995 might occur in 20% of years by the 2050s" (Hulme et al. 2002, p. 10). Extreme events like heavy precipitation, especially in winter, may be increase more than 20% which has a chance of occurrence from 50% of today. The extreme sea levels events, mainly caused by storm surges, might have the largest increase in the south-east coast of England. This region may experience also large changes in winds and storms and has the greatest decrease in height of the land.

Global climate change models have a resolution of 200 × 200 km, this resolution is sufficient for many global questions regarding climate change and possible consequences. The global climate change models do not directly provide adequate information about regional characteristics of climate change scenarios, but there results are useful as input parameter for regional downscaling models. In Germany, the Federal Environment Agency (Umweltbundesamt), develops climate change scenarios on the basis of two models: the REMO (Regional Model) of the Max Planck Institute for Meteorology and the WETTREG (Model for Regional Highly Resolved Weather Conditions) of the Climate and Environment Consulting Potsdam GmbH (CEC). These two models represent two different approaches of down-scaling: dynamically and statistically. The main difference between these approaches is that dynamic regional models use the results of the global model directly as input parameters and statistical models are based on statistical relationships between global climate patterns and regional weather conditions.

The climate change model REMO bases its calculations on a grid of 10 × 10 km; it provides information about climate change scenarios for Germany until 2100. WETTREG is also based on the emission scenarios (SRES) provided by the IPCC and by the results provided by the climate model ECHAM5/MPI-OM. Additionally, WETTREG uses input data from stationary climate and precipitation gauging stations.

Important for coastal protection issues is – besides the knowledge about precipitation with respect to the water management in the protected area – the development of storm surges and storminess. The possible development of storm surges in the southern North Sea was recently investigated within the PRUDENCE (*P*rediction of *R*egional Scenarios and *U*ncertainties for *D*efining *E*uropea*N* *C*limate Change Risks and *E*ffects) project of the Fifth Framework Programme of the EU (see prudence.dmi.dk, last visit: September 2007). The aim of this project was to provide detailed scenarios and deeper insight into the possible effects of climate change on regional level (Christensen et al. 2002). Several European institutions participated in this project. The German participants were the Institute for Coastal Research (GKSS) and the Max Planck Institute for Meteorology. The first one develops regional scenarios for extreme high water events at the North Sea coast. Scenarios for the time horizons 2030 and 2085 were calculated according to the scenarios of the PRUDENCE project. The scenarios were calculated on the basis of the global climate change models of the Hadley Center or the Max Planck Institute for Meteorology. Two emission scenarios (SRES) A2 and B2 were used as boundary conditions for the modelling. The input data from the global climate change models has been down-scaled to a grid of 10 × 10 km, which causes certain problems, e.g. the uncertainty of high water levels in future time (Grossmann et al. 2007).

Further information about scenarios for storm surge levels can be found in, e.g. von Storch (2006, 2007), Weisse and Plüß (2006), Weisse and Günther (2007), Woth et al. (2006).

4.3.2 Assessment Frameworks

The key to the selections of evaluation methods is provided by the preferences of individuals and organisations. The second basis is the maximisation of benefits. To quantify the preferences and the maximisation of benefit, it is mandatory to carry out an “economic appraisal” to evaluate costs and benefits of a project (Cost-Benefit Analysis, CBA). For the private sector and private companies this is crucial. Nowadays, this also applies to the public sector, because it is less accepted to spend tax on inefficient projects (Ruijgrok 2005). The difference between the private sector and the public sector is that cost and beneficial aspects are considered in a wider sense. That means the entire effects of the project should be considered, not limited to one organisation or institution. The application of economic appraisal methods is not standard procedure for coastal protection in Germany. Klaus and Schmidtke (1990) prepared a study for the Federal Ministry of Agriculture, Food and Forestry in Germany. Their assignment was twofold: First to conduct a cost-benefit analysis for coastal protection projects at the main land coast of Lower Saxony and second to develop a method which can be applied in practise. It was the first study which tried to evaluate the cost and benefit of coastal protection projects. Initial approaches to determining costs and benefits in water management have been developed in Germany – see e.g. LAWA (1981), DVWK (1985). Related investigations to coastal protection projects were: *Benefit Analysis of the Special Plan Coastal Protection for Sylt* by Klaus (1986), *Economic Appraisal for the Potential Flood-prone Area in Schleswig-Holstein* by Klug et al. (1998), *Sea Level Rise and Socio-Economic Consequences* by Behnen (2000) and recently, the investigation of *Micro Scale Evaluation of Risk of Flood-prone Coastal Zones* by Reese et al. (2002). The vulnerability of the county of Wesermarsch has been appraised to provide a sound basis for a database, which shall be used for the application of a CBA focussed on alternative coastal protection concepts (Kiese and Leineweber 2001). Only Klaus (1986) conducted a CBA, the other investigations focus on different items which contribute to CBA. For water management and coastal protection in Germany utility analysis and cost-effectiveness analysis were conducted to some extent, but were met with scepticism (Hartje et al. 2002). In Lower Saxony a CBA is not mandatory for the implementation of coastal protection projects as in the UK. Recently, multi-sectoral and integrated assessment approaches to local impacts of global climate change have been conducted in the UK for the East Anglia and North West England – see Holman et al. (2005a, b).

The fundamental outcome of CBA is to make evaluations transparent and fit for decisions (WBGU 1999) because most decisions in the real world are taken mainly on the basis of economic benefit. That means, that evaluation methods should contribute to demonstrating the relevance of non-economical aspects, e.g. biosphere or nature (WBGU 1999). Therefore, the aim of such methods is not to provide an exact figure, but to strive for the integration of all value categories as shown in Fig. 4.2. WBGU (1999) stressed that a focus should lie on the integration of economic evaluation, societal decision-making and the aims of sustainability. The application of a CBA in environmental projects is a main topic of current discussion – see e.g. Hartje et al. (2002), Hansjürgens (2004), Ruijgrok (2005), Convery (2007). The three main elements of criticism of CBA are the following (Hansjürgens 2004, p. 246):

- Criticism of the efficiency consideration as the underlying normative approach. This criticism is fundamental by nature and totally rejects CBA based on ethical considerations.
- Criticism of insufficient specifications of CBA. This criticism is directed at methodological shortcomings of CBA.
- Criticism of using CBA in the political decision-making process.

The first point concentrates on the weighting of criteria within a CBA. The weighting of criteria should discriminate one criterion against another, but decisions

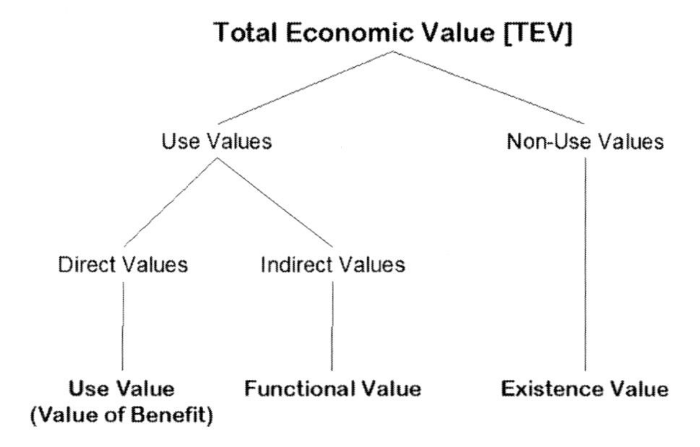

Fig. 4.2 Total economic value approach
Source: Adapted from WBGU (1999).

in the real world are done by weighting the considered criterion. The second one focusses on the uncertainty, the arbitrariness and the feigned accuracy of data. This is crucial part of the CBA. Strong effort has to be spent on the issue of data collection and data processing. There is no universal solution to these points but the applied method should be transparent regarding the implementation of data, the uncertainty and the accuracy. The last point of criticism focussed on the preference of politicians to decide on *hard* rather than on *soft* facts. Aspects which can be monetarised are seen as hard facts, the others as soft facts. Hansjürgens (2004) stated that the critique is related more to the incorrect use of the outcomes of a CBA not to the method itself.

The problem related to the issue of discounting is seen as a limitation to the application of CBA. Discounting should take cost and benefit of different time periods into account to enable a comparison (Hampicke 1991). Economists indicate that discounting should be applied for periods not longer then 10–20 years, because this period includes the time scale of one generation where the effects may be estimated, but to assess the effects which might occur for the next generations is impossible (regarding cost and benefit, e.g. Hampicke 1992, Endres and Holm-Müller 1998).

Conflicts may arise taking societal, economical and ecological aspects into account in a CBA. Knowing the different points of view, it is possible to anticipate potential conflict reasons by finding new solutions and compromises. Mainly, the applied methods of CBA try to monaterise goods whose value can not be determined by e.g. market prices. Monetary assessment is mainly done by willingness-to-pay methods. Different methods can be used to investigate how much customers are willing to pay, especially for non-directly quantifiable values (non-tangible). The advantages of willingness-to-pay methods are, that goods are related to a calculated value and that they can be integrated in a CBA. The main disadvantage of willingness-to-pay methods is that these values do not really represent the price people would pay for the goods – see e.g. Bateman (1995), Breidert (2005) and Guzman and Kolstad (2007). To estimate the influence of certain aspects of willingness-to-pay investigations, different methods can be applied. An overview of the integration of citizen's perspective in The Netherlands is given in Flinterman and Glasius-Meier (2005).

The approach for the case study Nessmersiel in Germany is based on the Total Economic Value (TEV) approach (Pearce and Turner 1990) which tries to cover all relevant value categories (see Fig. 4.2). "The approach considers different spatial dimensions (e.g. near-by or far away) as well as temporal dimensions (e.g. short-term or long-term) and it is possible to demonstrate the total amount of values which are related to the demand of natural goods. [...] The aim of this approach is not to allocate one calculated comprehensive figure, but to provide a discussion platform which considers all relevant value categories, e.g. benefit value, functional value, existence value, etc. (WBGU 1999). This comprises explicitly the ecological and social services of the biosphere, in our case especially applied at the pilot area Nessmersiel" (Meyerdirks and Ahlhorn 2007b, p. 6).

The total economic value distinguishes several value categories. The first step is to distinguish between use and non-use values. Direct (benefit value, symbol value), indirect (functional value) as well as future optional use (option value) are based on spatial proximity of goods, in our case the landscape itself. Non-use values (existence value) are based only on the knowledge about the goods, on whether the landscapes or animal species are near by or far away, without taking a direct benefit or using the services of the biosphere (Meyerdirks and Ahlhorn 2007b, p. 7).

Benefit value describes the direct use of services of the biosphere for production or consumption. The value of experience will also be subsumed under this value category, because it is related to consumption, i.e. tourism.

Functional value describes the indirect services of the biosphere. The ecological systems provide many regulation functions (e.g. water cycle, biogeochemical cycles or the composition of gases in the atmosphere) and structures (e.g. soil, dunes or slopes), which are necessary for human survival on the earth. Functions are usable characteristics of ecological structures. Values for the ecological goods consist of the utilisation of provided functions. The functional value describes the quality or the actual state of usable ecological services. For example: A wide fore land provide a regulation function for coastal protection, because it reduces to some extent the hydrodynamical load on the main dike, the soil provides a production function for agricultural use, and the existence of habitats, species and natural processes provide an information function for nature conservation.

Existence value is a non-use value. The existence value is neither based on direct nor indirect utilisation, only the knowledge of the existence of natural goods provide a value. Especially, the existence value is related to natural goods which will never be used, even in the future. To make the existence value tangible, especially for the German pilot area Nessmersiel, a questionnaire was applied to investigate the characteristics of the existence value of the participants for the Wadden Sea region. A detailed description is given in the box below.

For example, a comprehensive description of valuation of ecological functions is provided in de Groot et al. (2002). The requirements for the socio-economic-ecologic evaluation scheme which should be used and applied in the case study Nessmersiel are as follows: The method should

- integrate socio-economic as well as ecological aspects,
- enable to evaluate tangible as well as in-tangible aspects,
- integrate monetary as well as non-monetary values,
- be able to consider and to compare categorised criteria of different aspects (multi-criteria analysis),
- be able to compare different scenarios,
- be transparent and traceable.

Existence Value for the Pilot Area Nessmersiel

This short description of the determination of the existence value is a summary of the preparatory work for the one-to-one meetings and the consensus workshop of the pilot area Nessmersiel.

Within a technical report the intention of the existence value and the aim of its application at Nessmersiel was explained. This material was sent out several days in advance of the meetings, so that every participant could read the document and prepare himself. A hand-out was prepared with the aim, the criteria and a short description of the existence value for each integrated scenario. We conducted interviews about the existence value on the basis of this information to explore the completeness and the traceability of the description, the significance and the mode of payment for the existence value. The last point was not used to determine the characteristics of the existence value for the evaluation method, because all participants had agreed that a qualitative description of the characteristics of the existence value is sufficient. The participants were aware of the advantages as well as the disadvantages of willingness-to-pay methods. To explore the characteristics of the existence value for the pilot area Nessmersiel the criteria encompassed the landscape of the Wadden Sea Region, because the pilot area around Nessmersiel is a small section of the Wadden Sea Region in Lower Saxony.

The proposed criteria are as follows:

- typical plant and animal species (e.g. birds, salt marsh plants)
- natural habitat structures (e.g. salt marsh, beaches, tidal flat)
- natural characteristics of these structures (e.g. wideness of the landscape)
- natural process (e.g. erosion and sedimentation)

The participants extended the list with three additional criteria:

- cultural landscape, infrastructure
- homeland, attractiveness of the landscape
- climate change

The representative of a participating organisation was asked e.g. whether two people attending a one-to-one meeting would have to agree upon one vote. Question 4 asks the participants about the likelihood that the integrated scenarios will be implemented in 2050. The votes for the integrated scenario "A" are: 4 predominantly, 3 partly and 1 rather not. For the integrated scenario "B" the votes were 2 predominantly, 2 partly, 3 rather not and 1 never. For the integrated scenario "C" the votes were 2 predominantly, 4 partly, 1 rather not and 1 never. This shows that "A" might be the scenario which, in opinion of the group members, will be realised to some extent in 2050. The votes show that the group member prefer the implementation of "C" to "B". Question 5 asks for the characteristic of the existence value in each integrated scenario. The result is that "C" meets the defined criteria best, for "A" partly and for "B" rather not. This shows a preference of "C" to the others. The last two questions asked about the mode and the amount of payment for the maintenance of the existence value. Here, only the result of the mode of payment will be mentioned showing a preference for a voluntary mode of paying for the existence value, e.g. bounties, foundation, etc. Less preferred is the mode of entrance fee or tax.

4.3.3 Outranking of Scenarios – Background

In the previous section the intention and the limitation of a Cost-Benefit Analysis (CBA) was discussed briefly. Mainly, CBA are conducted by using a utility function which should be either maximised or minimised. Thus, the aim is to find an optimal solution for such problems. Most real-world problems are not related to one criterion, but to many criteria like social, economical and ecological. For example, for the interlinkage between multi-criteria analysis and sustainable development see Munda (2005). Applying a utility function, it is impossible e.g. to maximise all criteria. On the other hand, in real-world problems it is not wise to maximise all criteria, it is better to find a compromise. Therefore, a new category of multi-criteria decision aiding methods was developed: Outranking methods – see e.g. Brans and Vincke (1985), Roy (1985, 1996). For the German pilot area Nessmersiel the outranking method PROMETHEE (Preference Ranking Organisation Method for Enrichment Evaluation) was chosen (Brans et al. 1984, Brans and Vincke 1985).

4.3.3.1 Outranking Methods

The aim of outranking methods is to support the decision maker. Outranking methods do not focus on maximising or minimising an utility function, their intention is the pairwise comparison of one criterion with regard to two alternatives. Outranking methods use preference functions which indicate the preference, indifference or incomparability of criteria. A comprehensive overview about outranking methods and their characteristics can be found in Guitouni and Martel (1998) and a more general and comprehensive description of Multiple Criteria Decision Analysis (MCDA) can be found in Belton and Stewart (2002). Two outranking methods were developed in the earliest stage of MCDA and are widely used for various purposes: PROMETHEE and Elimination et Choice Translation Reality (ELECTRE). ELECTRE was applied to many problems to support decision makers, but has the disadvantages of requiring a lot of parameters. Some of these influence the results in a way which is not clearly understood and the application is more complex – for further explanation see e.g. Brans and Vincke (1985), Belton and Stewart (2002), Ruhland (2004). Salminen et al. (1998) compared three different outranking methods within the context of environmental problems and concluded that more than one method should be applied if possible. But, the differences between the outcomes of the applied methods are not great, they vary mainly in their complexity. Finally, Salminen et al. (1998) stated that PROMETHEE is easier to use than ELECTRE.

Guitouni and Martel (1998) provided seven tentative guidelines to choose an appropriate method for MCDA. For each guideline remarks are given concerning the application of the PROMETHEE method at Nessmersiel. The tentative guidelines are as follows (Guitouni and Martel 1998, p. 512):

Guideline 1 Determine the stakeholders of the decision process. If there are many decision makers, one should think about group decision making methods or group decision support systems. *Remark*: In Nessmersiel there are many decision makers, and the chosen method meets these requirements well.

Guideline 2 Consider decision maker 'cognition' when choosing a particular preference elucidation mode. *Remark*: The intention in the German pilot area was to compare three scenarios, so the elucidation mode is at least a comparison of criteria.

Guideline 3 Determine the decision problematic pursued by the decision maker. If the decision maker wants to get an alternatives ranking, then a ranking method is appropriate, and so on. *Remark*: The results of PROMETHEE will be a ranking of alternatives, i.e. scenarios.

Guideline 4 Choose the multi criterion aggregation procedure (MCAP) that can handle the input information available properly and for which the decision maker can easily provide the required information; the quality and the quantities of the information are major factors in the choice of the method. *Remark*: This is a very important aspect to be considered in conducting an evaluation process. PROMETHEE provides the necessary advantage and the required information can be provided by all participants easily.

Guideline 5 The compensation degree of the MCAP method is an important aspect to consider and to explain to the decision maker. If he refuses any compensation, then many MCAP can not be considered. *Remark*: The chosen method PROMETHEE is partially compensatory, some kind of compensation is accepted.

Guideline 6 The fundamental hypothesis of the chosen method are to be met (verified), otherwise one should choose another method. *Remark*: The fundamental hypothesis of the PROMETHEE method was met, thus it could be applied at Nessmersiel.

Guideline 7 The decision support system provided by the method is an important aspect which should be considered. *Remark*: In the German pilot area no decision support system was applied. The PROMETHEE method and the scoring matrix was implemented in a spreadsheet software provided by the author.

Bearing this in mind, the above mentioned guidelines should be met by the evaluation method, before the decision is taken to use an outranking method. The method should be able to handle different values and categories of criteria. Therefore, this kind of method belongs to the multi-criteria methods. Brans and Mareschal (2005) described in their article how the PROMETHEE method should support decision makers in taking their decisions. The method should overcome the disadvantage of many evaluation methods to calculate an optimal solution by using a utility function. "For this reason B. Roy (see Roy 1985, 1996) proposed to build outranking relations including only realistic enrichments of the dominance relation" (Brans and Mareschal 2005, p. 166).

4.3.3.2 PROMETHEE – Theoretical Background

This section describes the theoretical background of the PROMETHEE method. This description will act as the basis for the results of the Participatory Integrated Assessment (PIA) process generated in Sect. 5.2.5. The description is based on the articles of Brans and Mareschal (2005, pp. 164–175) and Esser (2001).

A multi criteria problem is given by:

$$\max\{g_1(a), g_2(a), \ldots, g_k(a) \mid a \in \mathbf{A}\}$$

where $\mathbf{A}$ is a finite set of possible alternatives $\{a_1, a_2, \ldots, a_n\}$ and $\{g_1(\cdot), g_2(\cdot), \ldots, g_k(\cdot)\}$ a set of evaluation criteria.

The objective of the outranking methods is not to find an optimal solution, but to find a compromise considering all (relevant) criteria. The outranking method PROMETHEE is based on the dominance relation.

Definition 1. (Dominance Relation) The principle of dominance is defined as: An alternative $a \in \mathbf{A}$ *dominates* an alternative $b \in \mathbf{A}$ iff for all $1 \leq j \leq k$ the inequality $g_j(a) \geq g_j(b)$ holds and for at least one k it is true. The principle of dominance is that an alternative a is preferred to b if b is dominated by a. An alternative is *efficient* if this alternative is not dominated by others. If indifference is defined as equality for all criteria then the principle of dominance is a *dominance relation*.

The dominance relation defines three preferences: For each $(a, b) \in \mathbf{A}$:

$$\begin{cases} \forall j: & g_j(a) \geq g_j(b) \\ \exists l: & g_l(a) > g_l(b) \Leftrightarrow aPb \end{cases}$$
$$\forall j: \quad g_j(a) = g_j(b) \Leftrightarrow aIb$$
$$\begin{cases} \exists s: & g_s(a) > g_s(b) \\ \exists r: & g_r(a) < g_r(b) \Leftrightarrow aRb \end{cases}$$

where P, I and R respectively are *preference*, *indifference* and *incomparability*.

To find a compromise each involved person has to specify different kinds of relationships: the specification how to determine the preference of two alternatives and the relevance of each criterion. The preference between alternatives is given by the applied preference function (comparison of each criterion of two alternatives). For the case study Nessmersiel the strong preference is applied:

Let P_j be the preference function for PROMETHEE with $P_j : \mathbf{A} \times \mathbf{A} \to [0, 1]$ which indicates the strength of the preference regarding one criterion. If $P_j(a, b) = 0$ there is no preference and if $P_j(a, b) = 1$ there is strong preference of a to b.

The relevance of each criterion is given by weights, which are non-negative numbers. It is not necessary to have normed weights, but to some extend these are easier to calculate:

$$\sum_{j=1}^{k} w_j = 1 \tag{4.1}$$

The outranking relation can be defined as $\pi : \mathbf{A} \times \mathbf{A} \to [0, 1]$, where π is the weighted mean of the preference function with

$$\pi(a, b) := \sum_{j=1}^{k} w_j P_j(a, b). \tag{4.2}$$

So, $\pi(a, b)$ expresses the degree to which a is preferred to b over all criteria, and $\pi(b, a)$ how b is preferred to a. For $a \in \mathbf{A}$ the out-flow $\Phi^+(a)$ and the in-flow $\Phi^-(a)$ are defined as follows:

$$\Phi^+(a) := \sum_{b \in \mathbf{A}} \pi(a, b), \tag{4.3}$$

$$\Phi^-(a) := \sum_{b \in \mathbf{A}} \pi(b, a). \tag{4.4}$$

The out-flow $\Phi^+(a)$ expresses how an alternative a outranks all the other alternatives. The higher $\Phi^+(a)$, the better the alternative. The in-flow expresses how an alternative a is outranked by all the others. The lower $\Phi^-(a)$, the better the alternative.

The partial ranking (PROMETHEE I) is obtained from the out-flow and the in-flow:

$$\begin{cases} aP^Ib \quad \text{iff} & \begin{cases} \Phi^+(a) > \Phi^+(b) \text{ and } \Phi^-(a) < \Phi^-(b) \text{ or} \\ \Phi^+(a) = \Phi^+(b) \text{ and } \Phi^-(a) < \Phi^-(b) \text{ or} \\ \Phi^+(a) > \Phi^+(b) \text{ and } \Phi^-(a) = \Phi^-(b); \end{cases} \\ aI^Ib \quad \text{iff} & \quad \Phi^+(a) = \Phi^+(b) \text{ and } \Phi^-(a) = \Phi^-(b); \\ aR^Ib \quad \text{iff} & \begin{cases} \Phi^+(a) > \Phi^+(b) \text{ and } \Phi^-(a) > \Phi^-(b) \text{ or} \\ \Phi^+(a) < \Phi^+(b) \text{ and } \Phi^-(a) < \Phi^-(b); \end{cases} \end{cases} \tag{4.5}$$

where P^I, I^I and R^I respectively mean *preference*, *indifference* and *incomparability*.

The difference of the out-flow and the in-flow is called complete ranking (PROMETHEE II):

$$\Phi(a) := \Phi^+(a) - \Phi^-(a) \tag{4.6}$$

with

$$\begin{cases} aP^Ib \quad \text{iff} & \Phi(a) > \Phi(b) \\ aI^Ib \quad \text{iff} & \Phi(a) = \Phi(b) \end{cases} \tag{4.7}$$

Alternative a is really preferred to b iff $\Phi(a) > \Phi(b)$ and is indifferent iff $\Phi(a) = \Phi(b)$.

If the following two assumptions hold for PROMETHEE, then the preference relation PROMETHEE I and PROMETHEE II are complete with regard to the dominance relation (see Definition, and Esser 2001, pp. 194–195):

1. All weights are definitely positive.
2. The preference relation to determine the preference of two alternatives should be positive or at least “0”. This assumption is no restriction for the case study Nessmersiel, because of the application of the strong preference relation.

That means that there is additional information available on which decisions and the comparison of alternatives will be made by the application of the outranking method PROMETHEE. Additionally, if for an alternative a the net-flow $\Phi(a)$ is positive and not dominated by another alternative, a is really efficient with regard to the above stated Definition on p. 114.

To determine the *profile* of one alternative, i.e. the performance of one alternative regarding each criterion, the net-flow has to be studied in more detail. Let:

$$\Phi(a) = \Phi^+(a) - \Phi^-(a) = \sum_{j=1}^{k} \sum_{x \in \mathbf{A}} [P_j(a,x) - P_j(x,a)] w_j. \tag{4.8}$$

This equation can be transformed to:

$$\Phi(a) = \sum_{j=1}^{k} \Phi_j(a) w_j \tag{4.9}$$

if

$$\Phi_j(a) = \sum_{x \in \mathbf{A}} [P_j(a,x) - P_j(x,a)]. \tag{4.10}$$

The practical application of the *profile* will be explained in Sect. 5.2.5.

The description of the theoretical background of the outranking method PROMETHEE substantiate the results for the pilot area Nessmersiel, i.e. the highest ranked alternative is really efficient against the others.

4.4 Cooperation for the Future – Participatory Integrated Assessment

The first part of this chapter deals specifically with the German situation with regard to spatial planning and coastal protection. The aim was to identify the gaps and bottle necks in the spatial planning system in order to implement spatial coastal protection concepts. The comprehensive investigation of Flügel and Dziatzko (2008) identified options and possibilities within the spatial plans and programmes at different level in Lower Saxony. The State Spatial Planning Programme (LROP) was amended over the last two years (ML 2008). The amended plan offers opportunities to implement spatial coastal protection concepts, because it concedes that climate change may impose further threats on coastal zones, especially to coastal protection, and therefore necessary space should be secured for new or alternative strategies. And "in consideration of climate change the research, development and testing of alternative coastal protection strategies should be considered" (ML 2008, point 1.4 number 12). Furthermore, necessary aspects have been incorporated into the amended LROP, e.g. the advice of stronger cooperation and enhanced participation and the extension of the area of validity to the 12 nm zone. Nevertheless, this advice has to be implemented within the Regional Spatial Planning Programmes (RROP), because the LROP will serve as a framework for the drafting of the RROP. Each RROP has its own approach because of the specific situation in each county. Some of these RROP have been amended over the last years, but in the light of the current LROP they have to be adapted again. The investigation of Flügel and Dziatzko (2008) shows that the perception of cooperation, participation and especially the consideration of risk differs. Whilst the State emphasises incorporating, as early as possible, the option for alternative coastal protection concepts, the counties will abide by the current concept of coastal protection. The opportunities of participation, i.e. involvement of the public during the development of plans and programmes, is not widely accepted. Some have relevant experiences and some not. The barriers to participation which have been identified in the second part of this chapter are valid. The experiences of the case studies conducted within the ComCoast project produced nine barriers (see Sect. 4.2.2). Furthermore, solutions have been suggested to overcome these barriers, which might emerge before a project starts, but which might also occur during a project. Nevertheless, several methods and techniques are available to tackle these problems. The choice of the right method is complicated, but this process has to be completed on time and carefully to deliver a good and well-founded participation process. Several guidelines for the identification of the right method exist and the available experience shows that each process has to be tailor-made. The demand for cooperation and participation within the spatial planning instruments in Lower Saxony led to the question: which method should be applied? In Germany, participation is neither applied in coastal protection projects nor in the compilation of the Master Plan; it is mainly used as a method for conflict resolution rather than conflict prevention or for exploring opportunities.

For example, concerning the fore land, there is no conflict whether a fore land is desirable or not, the conflict is about how the fore land should be treated e.g. to cope with erosion – see e.g. Kunz (1999a). In front of a main dike without a fore land sedimentation fields are built up. The sedimentation fields function to support the accretion of sediments and create flats – e.g. Erchinger (1970). The result is that under certain boundary conditions a salt marsh can grow. Consequently, the mud flat will be reduced and the salt marsh area increases. Thus, the National Park administration has to balance between salt marshes and mud flats (this remark was made at the international workshop of the ComCoast project in The Netherlands, held in September 2006, in respect of land reclamation work in front of a main dike). The decrease of fine-grained material – see e.g. Michaelis (1968), Meyer and Ragutzki (1999), Mai and Bartholomä (2000), Eppel and Ahrendt (2005) – may lead to a discussion about the occurrence and the importance of different natural units within the Wadden Sea. Recently, these findings were substantiated by a PhD-thesis written at the University of Kiel on morphodynamics and habitat changes in the Wadden Sea, which concludes that a progressive coastal squeeze of the Wadden Sea is likely and consequently a significant change in grain-size will occur. This directly influences habitats of the Wadden Sea, especially sea grass and mussel beds (Dolch 2008).

For a detailed description of conflicts see e.g. Kunz (1991, 1994), Ahlhorn (1997), Ahlhorn and Kunz (2002a) and Kunz (2004a).

Nowadays, not only the involvement of the broad public is required; it also necessary to apply a sound economic appraisal to projects. For the private sector this is crucial. This is also valid for the public sector, because there is an increasing awareness of cost-benefit, which also reflects possible impacts on the environment (see Sect. 4.3.2). The difference between the private and the public sector is that cost and beneficial aspects have to be considered in a wider sense. That means that the total effects of projects should be considered, and assessments not be limited to one organisation or institution. The problems of cost-benefit-analysis have been described in the third part of this chapter. The main problem is that the identification or allocation of one figure to everything is impossible, i.e. monetarisation of all goods. Methods are available to monetarise natural goods, e.g. willingness-to-pay, but these methods have inherent problems. The estimated figure for a natural good might not really express the figure which would be paid by individuals. It is complicated to determine prices for the natural environment, like salt marshes or birds. To overcome these problems, the application of descriptive categories can be used. But, the applied socio-economic evaluation method should be able to handle these categories. Thus, traditional CBA methods using a utilisation function can not applied offhand (see Sect. 4.3.2). Therefore, outranking methods have been developed to overcome this shortcoming of traditional CBA methods. The advantage of outranking methods is that figures as well as verbal categories can be handled. Furthermore, the purpose of outranking methods is to support the decision maker, and to structure the available information within the decision-making process. Also, the theoretical background has shown that with additional assumptions, which do not restrict the applied PROMETHEE method, will enhance the outcome of the outranking method (see Sect. 4.3.3). So, the ranking method PROMETHEE, is really equipped to deliver an efficient alternative for solving a problem.

Within planning processes, information is needed on the one hand on "alternatives" and on the other hand on "criteria". The information on alternatives is e.g. given by preference functions, and the information on criteria is mainly given by weights. Within the compilation of a spatial planning programme the comments and remarks of other institutions, organisations or even the broad public have to be considered. But, there is no general consents how this can be done. How is each remark or comment to be weighted? Which have to be taken into account? Who decides which aspect is more important than another?

The discussion of available instruments (from spatial planning and coastal protection) in this section indicate that these instruments provide necessary structures. The existing plans and programmes for spatial planning offer the opportunity to implement spatial coastal protection concepts. In many documents, in spatial planning as well as in e.g. the national strategy on ICZM, the involvement of relevant stakeholders, target groups and organisations is requested, but there is no proposal on how to conduct participation. Participation is carried out in a mainly formal way, and this is restricted to commenting on plans and programmes, but does not foresee direct joint development of these plans and programmes. It is strongly recommended to consider other sectors and organisations, when developing a plan or programme.

4.4.1 Framework for Implementation

All pilot projects of the ComCoast project were run in a similar way (see Chap. B): The experiences and the knowledge of different pilot projects led to a framework of implementation (see Fig. 4.3). Feasible and sound options will be the results

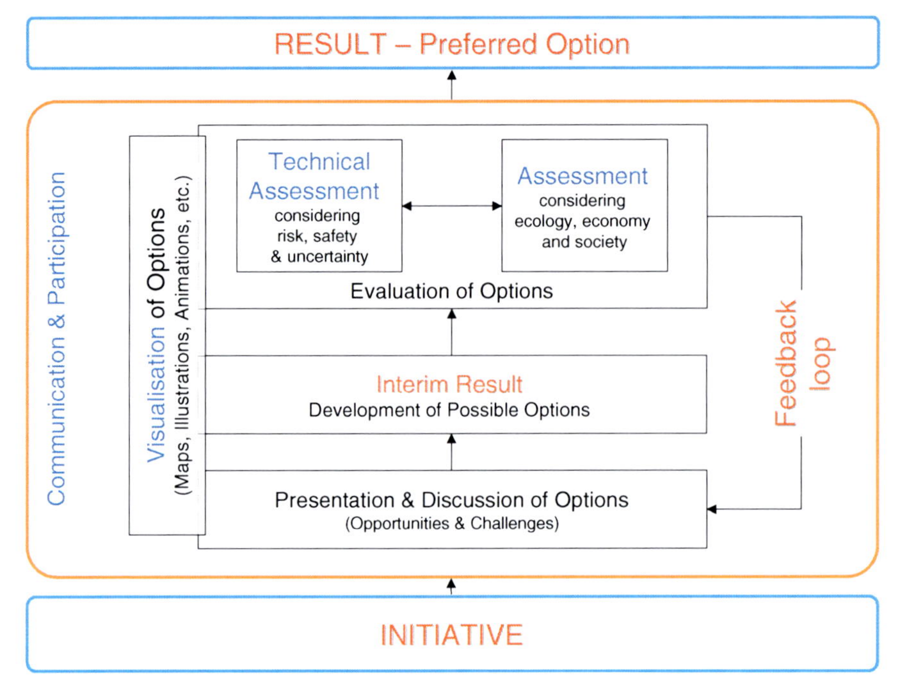

Fig. 4.3 Framework for implementation
Source: Meyerdirks and Ahlhorn (2008).

of the entire process. The Participatory Integrated Assessment (PIA) process (see Sect. 5.2) in Nessmersiel was a special case in the whole context of implementing Multifunctional Coastal Protection Zones (MCPZ, see Sect. 5.1) around the southern North Sea – e.g. RWS (2007a).

The framework, shown in Fig. 4.3, provides a general guideline and serves as a flow-chart for the implementation of spatial coastal protection concepts (Meyerdirks and Ahlhorn 2008). Each process starts with an *Initiative*. For example, the strategic aim may be: a beautiful, prosperous and safe coast. But, how to achieve this?

- Apply participation within the entire process.
- Visualise the opportunities of the proposed options.
- Design and evaluate different (spatial) coastal protection concepts.
- Evaluate the proposed options considering social, economic and ecologic aspects.

After the initial step the proposed options should be presented and comprehensively explained to a selected group. The composition of this group is the result of a thorough stakeholder and force-field analysis – e.g. Stroobandt et al. (2007). Within this group the challenges and opportunities of the proposed options should be discussed and elaborated. Selected options should be investigated from the technical point of view, i.e. conduct a technical design and assessment process. After this technical assessment the integrated assessment of social, economic and ecologic aspects should be conducted. The results of these assessments should be visualised in a transparent and traceable manner. The feedback-loop will assure that all options and aspects have been considered. If the entire group decides upon a feasible solution (i.e. preferred option), then the process can be concluded. The methods and tools to be used to run this process are tailor-made and depend on the contents and the context of the *Initiative*.

The open points will be tackled within the PIA process conducted in the case study Nessmersiel. It will be shown that the combination of participation and socio-economic and ecologic evaluation can lead to a jointly (in this case limited to responsible institutions) accepted planning process and result, namely sustainable coastal development.

Chapter 5
Sound Options – Multifunctional Coastal Protection Zones

Contents

5.1 Spatial Coastal Protection Concept

5.1.1 Definition of Coastal Protection Zones

In the previous section, the rationale behind the coastal protection system and the strategies in force have been dealt with. The conclusion is that new or revised concepts and strategies are necessary to meet the challenges of the future. However, some problems demand immediate attention and the consequences of climate change will make them even worse.

However, coastal protection is not the only user perspective at the coast, but for the low-lying areas in Lower Saxony an important one. The previous sections on spatial planning and ICZM have shown that other user perspectives have to be considered in coastal planning. This is especially the case if the approach of widening a single line of defence to a defence zone will be implemented.

This section is based on a working definition of (Multifunctional) Coastal Protection Zones (MCPZ). Bosecke (2005) described the approach of the former German Democratic Republic; namely the installation of coastal protection areas

F. Ahlhorn, *Long-Term Perspective in Coastal Zone Development*,
DOI 10.1007/978-3-642-01774-2_5, © Springer-Verlag Berlin Heidelberg 2009

(Küstenschutzgebiete) and coastal protection stripes (Küstenschutzstreifen) in the State Mecklenburg-Vorpommern (still valid). The coastal protection area is determined by the flood-prone area, i.e. the area lying below a determined contour-line. The coastal protection stripe is about 200 m at the Baltic Sea and about 100 m at the Bodden coast. The installation of the coastal protection area should guarantee a continuous protected strip along the coast of Mecklenburg-Vorpommern. The intention of the coastal protection strip is to prevent further housing and development. Within the coastal protection areas all development and e.g. forest management projects require the permission of the responsible water management authority. The main intention of both was to prevent further development and to minimise the accompanied increase of risk and to ensure enough space for coastal protection in the future. Bosecke (2005) emphasises these approaches under the changing circumstances caused by climate change and other threats and trends (as a prominent role). Therefore, the definition of MCPZ refers to this approach, taking the specific situation of Lower Saxony into account, i.e. the State Law on Dikes.

In Fig. 5.1 the proposed territorial definition of coastal protection territory, zones and strips is shown.

A Coastal Protection Strip (CPS) already exists within the State Law on Dikes (NDG, MU 2004): §16 sets a 50 m buffer on the landward side of the main dike which has to be kept free of constructions. On the seaside, §21 in combination with §23, demands for a sufficient width of the fore land: If the width of the fore land is less than 200 m, then a buffer of 500 m of the Wadden Sea can be designated for

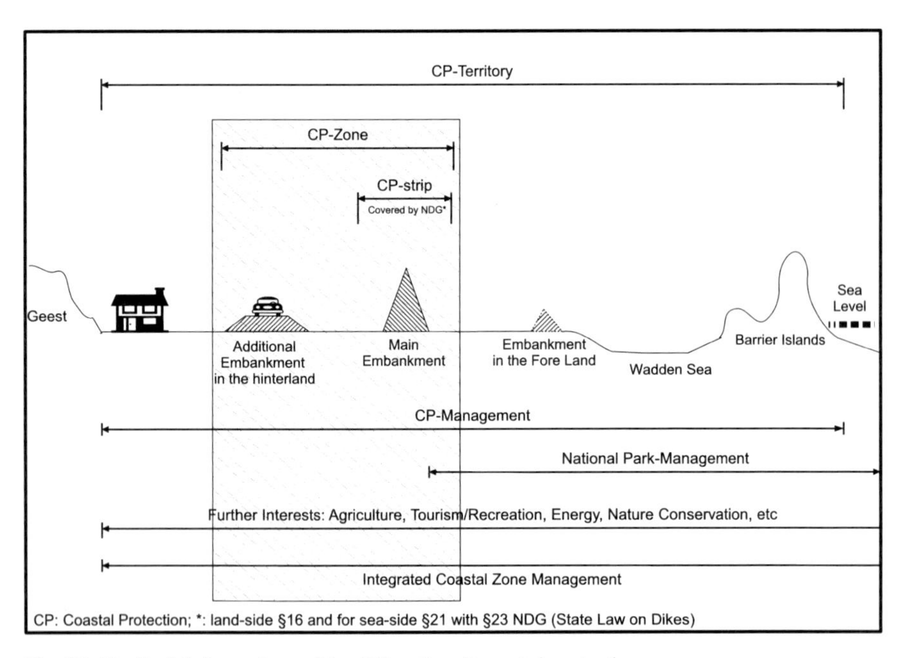

Fig. 5.1 Territorial dimensions of (multifunctional) coastal protection zones

coastal protection as a safety buffer (Sicherungsstreifen). Thus, the coastal protection strip ranges from 50 m landward to max. 500 m seaward. With regard to the effects of the heightening of existing dikes, demand for more space, i.e. widening of the dike foot, the enhancement of this strip should be considered. For example, in Schleswig-Holstein the distance between the border of the Wadden Sea National Park and the dike is about 150 m (§3 Law on National Park Wadden Sea Schleswig-Holstein, NPG-SH 1996). The diverging approaches to maintain and preserve the fore land and the introduction of new objectives within the Law on Nature Conservation in Schleswig-Holstein led to the installation of a working group to deal with the complex of problems which arose – see MELFF (1995) and Hofstede and Schirmacher (1996). The positive character of the fore land management plan was approved by a report on the experiences gathered within the first five years of the plan's enforcement (MLUR 2001). In Lower Saxony, first experiences have been made with the preparation of fore land management plan for the dike board Norden (NLWK Norden 2003).

The next level is the Coastal Protection Zone (CPZ). This area encompasses the seaside buffer of the CPS and stretches out to an additional embankment in the hinterland, which is capable of protecting the hinterland in case of a main dike failure in the event of flooding or inundation, e.g. second or old dike line or elevated road. At present, the CPZ only exists on some coastal stretches along the Lower Saxonian coast, e.g. East Frisia (see Fig. 5.5a). These zones fulfil the recommendations of the Engineering Committee Coastal Protection (Engineering Committee for North and Baltic Sea 1962) based on the experiences of 1962 to install, where possible, second dike lines or, more generally, additional embankments behind the main dike.

The rationale for the definition has already been published and excerpts are quoted here: "The principle of the State Law on Dikes of Lower Saxony (NDG) is that all people and properties will be protected behind the dikes and therefore all people living in the protected area are members of a dike board". That is laid down by §6 paragraph 1 NDG: "... the owners of properties behind the dike (protected area) are obliged to ensure the common maintenance of the dikes". According to §9 paragraph 2 and 3 NDG the counties determine the landward border of the protected area which is the design water level. The border in the hinterland is normally the NN +5 m contour line, but for some dike boards it can be the NN +6 m or up to NN +8 m contour line (MU 2006). The safety standard that is given by the design water level and the local design wave run-up (§4 NDG) shall provide the same safety standard along the whole coast of Lower Saxony regardless of whether farm land, villages or cities are affected. Thus, the selection of indicators has to be made very carefully. Coastal protection is a high priority land use on the coast in Lower Saxony. Only with a fully functional coastal protection system that anticipates future challenges people can live and work in the floodplain of Lower Saxony. The selection of indicators is based on the State Law on Dikes (NDG). In §21 paragraph 1 NDG it is recognised that the "... area lying between the main dike and the mean high water line with or without a dike (fore land) ... has to be maintained as protection for the main dike by the dike board at an appropriate width". Indeed, the commentary to the State Law on Dikes (Lüders and Leis 1964) and §23 NDG propose a width of 200 m. Thus, one indicator is the existence of a broad fore land of at least 200 m. Regarding the hinterland, §29 NDG mentions a second dike line: "Dikes which are appropriate to give additional safety in the case of a breach in the main dike or failure of a barrier, have to be dedicated as second dike lines. [...] Thus, a second indicator will be the existence of a second dike line in the hinterland, regardless of whether it is formally dedicated or not" (Ahlhorn et al. 2007b, pp. 62–63).

The installation of a CPZ in the coastal zone of Lower Saxony is feasible, since the State Spatial Planning Programme (LROP) states in paragraph 1.4, section 12 (ML 2008): "Considering the consequences of climate change research, development and testing of alternative coastal protection strategies have to be taken into account." Thus, this stipulation allows for the establishment of CPZs in Regional Spatial Planning Programmes (RROP) as pilot areas. This requires the installation of a regional association of all coastal communities of Lower Saxony, which can develop a Regional Development Concept for the coast of Lower Saxony based on the Spatial Planning Concept for the Coast of Lower Saxony (ROKK). This Regional Development Concept should include the installation of CPZs for research, development and testing aspects (Flügel and Dziatzko 2008). Furthermore, the CPZ could be established as a priority area for coastal protection (see box in Sect. 4.1.1). Within this framework further types of land use need to be considered in line with the overall aim of sustainable coastal development.

The next level is covered by the Coastal Protection Territory (CPT), which ranges from the −15 m depth-line to the landward border of the dike boards. The landward side is covered by the NDG. The seaside border is defined through the influences of hydrodynamic and morphodynamic changes on the entire Wadden Sea system. The barrier islands are part of this system, but have to be treated in a different way. This is not within the focus of this dissertation.

The main conflicts occur from parallel demands made on the same space. Hence, the MCPZ provides various functions based on existing natural units like salt marshes and tidal areas on the seaward side and e.g. salt marshes on the landward side, for different user perspectives, e.g. tourism, industry, harbour, nature conservation and coastal protection. For example, Ahlhorn (1997) described inter- and intra-sectoral conflicts from parallel demands made on the same space. These conflicts have been discussed for the utilisation of the barrier islands with regard to tourism/recreation and drinking water management and the National Park. However, conflicts arise between and within sectors. Different kinds of land use rely on the same natural units. Mutual interests can cause problems if there are divergent aims – with respect to coastal protection and nature conservation see e.g. Petersen (1998), Ahlhorn and Kunz (2002a).

Does the territory definition provide the necessary basis for the integration of coastal protection into spatial planning? Starting with the Coastal Protection Strip (CPS) the interests and needs of coastal protection are clearly defined and accepted. There are conflicts on the sea side, e.g. with the National Park (see Sect. 4.4), but these conflicts focus mainly on *how* and not the *whether* a fore land be protected. The Coastal Protection Zone (CPZ), if already existing, may be designated in the future as a *priority area* for coastal protection with special regulations for other types of land use. The integration of other types of land use needs to be regulated. The Coastal Protection Territory (CPT) or protected area could be designated as a *precautionary area* for coastal protection. This can serve as an option for the long-term perspective, because changes in e.g. climate, demography and economy have to be continuously taken into account.

What about the coastal stretches where no CPZ exists? As already mentioned and recommended, a wider zone to protect people and their property will provide an added value under changing circumstances. Every project or initiative in the coastal zone, especially within a certain area near the high water line, should seriously consider the advantage of generating a CPZ (multifunctionality). In some places this is difficult for specific reasons and other solutions are needed. For example, the planning process for a motorway in the north-western part of Lower Saxony, the so called A22, is intended to the flow of traffic generated by the harbour areas between western Europe to Scandinavia and to eastern Europe. This motorway, however, might have an additional and beneficial effect on the safety of the hinterland in the north-western part of Lower Saxony, if it is built as an elevated road, to act like an embankment (second dike). This may serve only as an example for multifunctional use and should stimulate the development of multiple use concepts in different projects, thereby providing a step forward towards sustainable and integrated planning. For example, north of Wilhelmshaven, roads have been built on the older dike lines (former main dikes) and these dike lines serve as additional protection elements in the case of main dike failure. The infrastructure existing between the main and the older dike line has to be adapted to the special situation. Currently, this is not felt necessary, because the main dike is rated as high enough and sufficient. However, in the long-term this may serve as an option. Another possibility might be to revitalise individual former dike lines which to some extent are still visible. For example, on the eastern part of the Jade Bay, in Augustgroden, the road was built on an old dike line. This might, as long-term perspective, serve as additional embankment for the recently improved main dike. The area between the main dike and road can be used e.g. as grass or crop land, but the heightening of the main dike in the future might not be necessary, if the inner slope is strengthened by solutions proposed in the ComCoast project (e.g. geo-grid) to be made more overtopping resistant. This might have several benefits, e.g. less demand for material, less demand for space if the heightening (Fig. 3.19) is not necessary.

As a basis for cooperation between the National Park government and coastal protection the report on *Opportunities for compensation of coastal protection projects within the National Park* (BR W-E and NLPV 2001) could act as a starting point. This report comprises detailed descriptions of possible compensation measures for coastal protection. For the implementation of MCPZ these suggestions have to be considered.

Another example are the suggestions described by Jeschke (2004) for the coastal stretch from Varel (Lower Saxony, southern Jade Bay) to the eastern part of the Jade Bay. Different options were suggested to improve the currently poor condition of the main dike, which has to be strengthened in the coming years – see also Klenke et al. (2006).

The following sections address the question of *how* a (Multifunctional) Coastal Protection Zone (MCPZ) can be implemented with the consideration of both safety and the interests of other types of land use. The Participatory Integrated Assessment (PIA) process offers opportunities for both a sustainable coastal development and the strengthening of the safety level for coastal protection.

5.1.2 Proposed Solutions of the ComCoast Project

The present status of coastal protection can be illustrated as in Fig 5.2a. Germany and parts of The Netherlands and Denmark are protected against flooding by main dikes. Parts of Denmark and The Netherlands are also protected by dunes. The following illustrations show the solutions proposed by the ComCoast project. For the purpose of this dissertation a focus on solutions for low-lying areas (Figs. 5.2b, 5.3 and 5.4) has been selected.

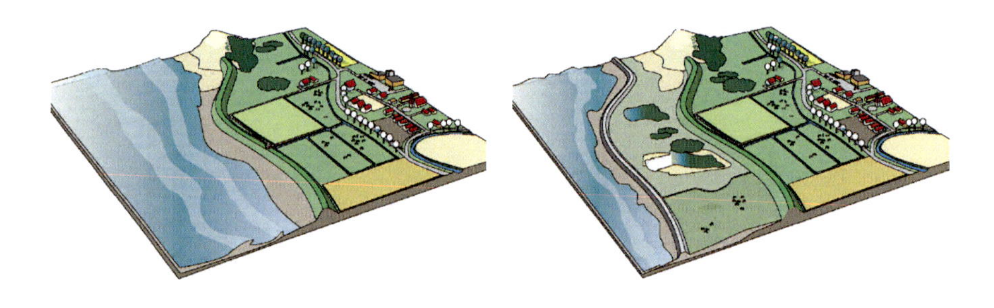

(a) Typical situation in low-lying areas (b) Illustration of Fore Shore Protection

Fig. 5.2 Typical situation in low-lying areas and fore shore protection
Source: © van Lint vormgeving.

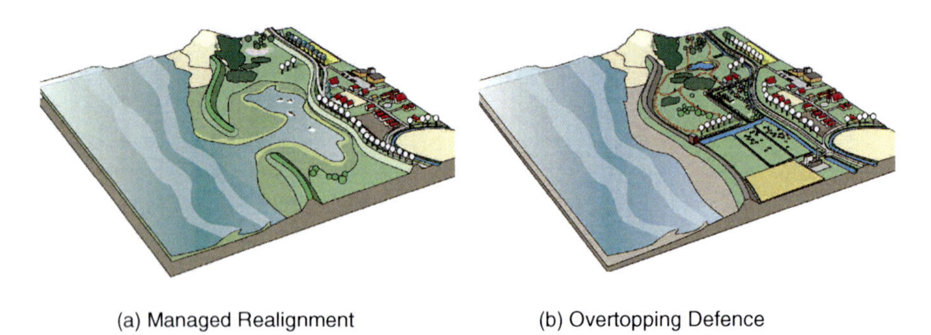

(a) Managed Realignment (b) Overtopping Defence

Fig. 5.3 Illustration for managed realignment and overtopping defence
Source: © van Lint vormgeving.

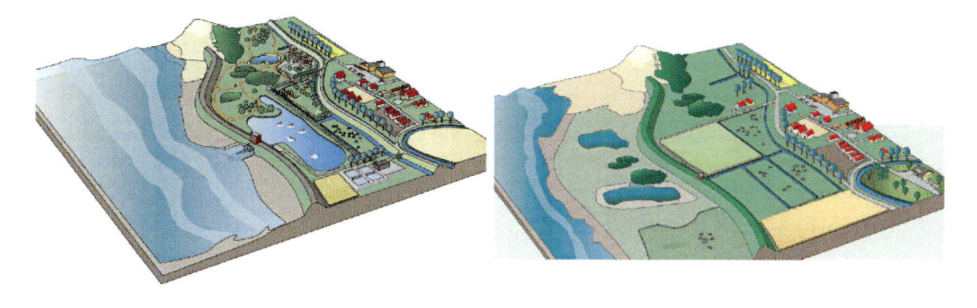

(a) Regulated Tidal Exchange (b) Fore shore Recharge

Fig. 5.4 Illustration of regulated tidal exchange and fore shore recharge
Source: © van Lint vormgeving.

The implementation of these proposed solutions depend on the given natural situation (e.g. tendency of erosion etc.) at the coast and the legally and statutory rights in the country in question. These proposed solutions are not new developments of the European ComCoast project. The new goal is to widen the single line of defence to a zone in which other functions are integrated, i.e. taking the multifunctional approach into account from the outset, when implementing a coastal protection project or a coastal development project. The latest investigation of the effects of these proposed solutions was made by INFRAM (2007). The INFRAM investigation shows that the most effective solution is to increase the roughness of the outer slope of the dike or flatten the profile of the main dike. The latter recommendation has already been applied in Lower Saxony where the outer slope of the main dike start at a gradient of 1:10 (1:12) and ends at the crest at a ratio of 1:6. Nevertheless, the testing of the overtopping resistance of a dike in Delfzijl (NL) indicates that volumes of water up to 50 l/s/m can be withstood without serious damages to the inner slope (RWS 2007b).

5.2 Results of the Case Study Nessmersiel – Achieving Multifunctionality

The main purposes of the PIA process in Nessmersiel were:

- To develop a participation process and to integrate an adequate evaluation method,
- to stimulate multifunctional thinking to achieve sustainable coastal development,
- to display options for spatial coastal protection concepts.

5.2.1 Identification of Sites

One objective of the ComCoast project was to identify feasible sites for the implementation of the spatial coastal protection concept. For that purpose, a phase model was developed, which provides guidelines on identifying feasible sites (Ahlhorn et al. 2007a). The experience of this European project shows that considerably different reasons may lead to the application of spatial coastal protection concepts (see Sect. 4.4). Besides that, within the partner countries different approaches have been applied to identify feasible sites – see e.g. Halcrow (1998a, b), Thomas (2002), ABPmer (2005), DHV (2005), Oedekerk (2006). Roughly, a minimum of basic requirements, shown in Table 5.1, can be provided as a common baseline. But, these basic requirements may be subject to discussion for specific reasons. For example, the installation of fore shore recharge may be executed under unfavourable hydrodynamic conditions, due to a high economic value of the protected area. The application of managed realignment may not be feasible, even if there is an additional embankment in the hinterland, due to negative attitudes towards releasing land to the sea. Several aspects have to be considered when looking for a feasible site, and not only technical aspects.

The following description is a preparatory step for the identification of feasible sites. This step will be elaborated for the German situation. The description is focussed on the classification of the Lower Saxonian coastal zone, using the definition of "coastal protection zone" (Sect. 5.1.1 and Fig. 5.1). Regarding this definition the Lower Saxonian coast can be classified as follows:

High: Existing Coastal Protection Zone (CPZ), i.e. fore land more than 200 m wide with second embankment.

Medium: The potential for a CPZ is given, i.e. either a second embankment or a broad fore land exists.

Low: Coastal Protection Strip (CPS) exists, with minor potential for a CPZ, i.e. a main dike exists.

In Fig. 5.5 the results of the investigation are shown: The green colour indicates a broad fore land as well as a second embankment a feature of the region of East Frisia, consisting of the old dikes from the last centuries (Fig. 5.5b). Yellow indicates the existence of either a broad fore land or a second embankment; in Lower Saxony

Table 5.1 Basic requirements for the proposed solutions for the implementation of multifunctional coastal protection zones

Proposed solution	Basic requirement
Regulated tidal exchange (RTE)	Appropriate elevation and second embankment in the hinterland or rising hinterland
Managed realignment (MR)	Appropriate elevation and second embankment in the hinterland or rising hinterland
Overtopping defence (OD)	Second embankment in the hinterland or rising hinterland
Fore land protection (FP)	Appropriate width of the fore land and appropriate hydrodynamic conditions
Fore shore recharge (FR)	Appropriate hydrodynamic conditions

Source: Ahlhorn et al. (2007a).

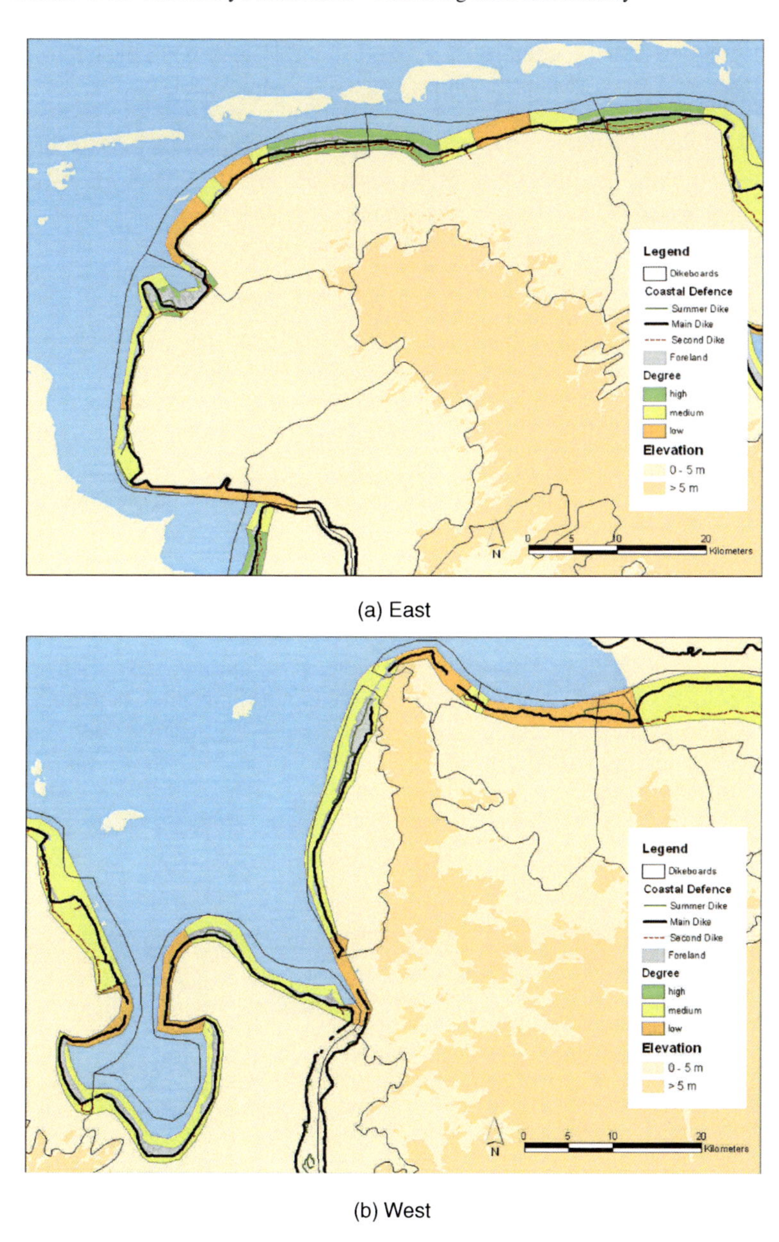

Fig. 5.5 Characteristic of the Lower Saxonian coast with regard to coastal protection
Source: Ahlhorn et al. (2007a).

these stretches generally exhibit a broad fore land. The orange coloured coastal stretches indicate the coastal protection strip on the main dike.

Finally, there are only few coastal stretches in Lower Saxony where a CPZ exist. But, there are many coastal stretches, which offer the opportunity of enhancement to a CPZ under certain circumstances, because either a second embankment or a broad fore land exists. The enhancement of these potential areas to CPZ has further to be investigated, because the specific circumstances of these sites have not been adequately considered in this first step. For example, assume that the potential area has a second embankment, but no broad fore land. Then e.g. the hydrodynamical conditions have to be investigated and it must be determined whether a broad fore land could be gained. In this case the cooperation with the management of the National Park would be required. If there is a broad fore land, but an additional embankment in the hinterland is missing, the enhancement to a CPZ depends on the configuration of the hinterland. For the latter case, the PIA process described in the following sections can serve as platform to enhance these sites.

The following description of the Nessmersiel investigation area shows that a CPZ is already in place. This CPZ consists of a broad fore land, a summer dike, a summer polder, a main dike and a second dike line. Nevertheless, the process can be applied to sites without these features with a view to activating their full potential.

5.2.2 Description of the Site

The area of the polder adjacent to Nessmersiel covers approximately 1,200 ha. The summer polder is about 237 ha and the polder behind the main dike is approximately 455 ha (Fig. 5.6). The area has a slight slope in north-easterly direction. Furthermore, the summer polders are situated 25 cm up to 50 cm above the elevation of the adjacent polders.

Tourist activities are restricted to cycling and walking along the dike's. The recreational zone of the National Park is located adjacent to the ferry harbour featuring

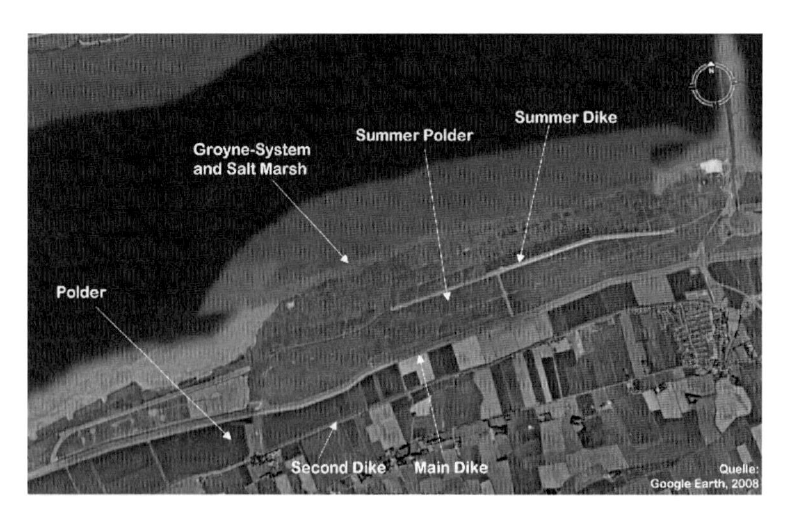

Fig. 5.6 Pilot area Nessmersiel in north-western part of Lower Saxony
Source: Adapted from Google Earth (2008).

(a) The restaurant at the ferry harbour of Nessmersiel

(b) The little beach at the ferry harbour

Fig. 5.7 Photographs of the pilot area Nessmersiel
Source: Ahlhorn (1997, 2005) @ Frank Ahlhorn.

tourist infrastructure like a restaurant and a small beach (Fig. 5.7). In the western part of the pilot area there is a path featuring directly to the Wadden Sea. The accessibility of the remaining area of the summer polder is limited due to the Wadden Sea National Park.

The polder areas behind the main dike are intensively used by agriculture, mainly as crop land (Fig. 5.8a). The fields can be accessed via crossings over the second dike. The summer polder is mainly used for grazing as part of the integrated fore land management plan – see NLWK Norden (2003). A maximum of 1.5 cattle per ha are allowed. On the summer dike 2.5 cows per ha are allowed. The main dike is grazed by sheep for dike maintenance (Fig. 5.8b).

The fore land has been designated as a restricted zone within the National Park. The summer polder is a significant salt marsh habitat. In contrast to the summer

(a) Photograph of the polder in June 2007

(b) Photograph of the summer polder in June 2007

Fig. 5.8 Photographs of the summer polder and the polder
Source: Ahlhorn (1997, 2005) @ Frank Ahlhorn.

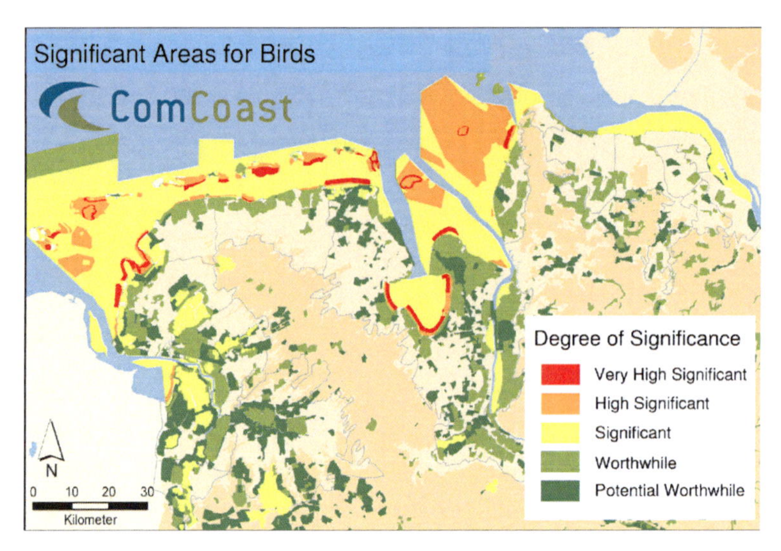

Fig. 5.9 Result of the aggregation of different levels of conservation and protection status' for birds Source: Ahlhorn et al. (2007a).

polder, the polder areas behind the main dike have a low conservation status. But, these polder areas are important resting and feeding areas for birds (see Fig. 5.9).

The main dike was strengthened in the middle of the 1980s. Within the pilot area, the main dike has been constructed around 300 m up to 700 m seawards. In the Master Plan for Coastal Protection in 1973, a height of NN +710 m was set as the design height for the main dike. A laser scan survey in 2004 indicated crest heights between NN +780 m and NN +830 m (NLWKN 2007a). The ratio of the outer slope is 1:6 and the ratio of the inner slope is 1:3. The summer dike was strengthened in the same period as the main dike. The ratio of the inner as well as the outer slope of the summer dike is 1:7. The summer dike crest height varies between NN +310 m and NN +330 m. The total length of the summer dike is approximately 9 km. The crest of the second dike line has a constant height of NN +500 m. The total length is 10 km, the second dike line is the former main dike built in the late middle ages.

In Nessmersiel, the main dike complies with the safety requirements of the Master Plan (Fig. 3.16). In the pilot area the second dike line is in bad condition in some stretches, because of cattle on the dike (Fig. 2.9b). The maintenance costs for the second dike line amounted to about 20,000 € in 2006; the summer dike in the pilot area is maintained by the State which costs about 82,000 € in 2006 (personal communication Mr. Manßen, GLL Norden). This can amount to even more depending on the amount of flotsam which occurs in the storm surge season.

5.2.3 Design of the Participation Process

Remark: Information about participation in the German pilot project has partly published in Ahlhorn and Klenke (2006a), some components are new due to the ongoing development of the participation process. Sections which have already been published are marked by quotation marks.

The Participatory Integrated Assessment (PIA) process was conducted on the following working hypotheses:

- It is possible to adapt the existing coastal protection system to the needs and interests of different user perspectives and assure at least the same level of safety.
- It is possible to create a planning process on the basis of the EU ICZM recommendations.

In Fig. 5.10 the flow-diagram of the PIA process is shown. The process was divided in one preparation step and three major steps: (1) kick-off meeting, (2) a consensus workshop on the land use pattern in the light of given scenarios for 2050, (3) consensus workshop on the weighting of criteria and (4) a final workshop for participatory assessment of the results and the entire process.

A comprehensive stakeholder analysis was conducted and the relevant stakeholders were identified: coastal protection authority, nature conservation organisation, dike board, chamber of agriculture, National Park management, spatial planning officers at regional and State level as well as representatives from the municipality and the tourist association. Because of the fictitious nature of the enterprise (planning exercise) and the awareness of the sensitive issues involved it was decided to execute an institutional participation process. A selected number of participants was invited to set up a short-term joint working basis.

The process is a mixture of elements of participation and evaluation. The structure of the process is based on the involvement and the progressive integration of all participants. All participants were directly involved and they had direct influence on the results at each step of the process. Intermediate steps were necessary to prepare workshops. The period in between two workshops was used to intensify the exchange of knowledge between the participants and the leading organisation (feedback loops). And, more importantly, the leading organisation was always accessible for concerns of participants and responded quickly to requests.

The purpose of the kick-off meeting was to become acquainted with the invited representatives and to introduce the intention and the objectives of the PIA process. Within the kick-off meeting information was gathered about the legal status of the pilot area, to provide a clear starting point for every participant. Therefore, each participant was asked to explain the needs, interests and legal status of the area from the point of view of his institution or organisation. Afterwards, the basic settings for the year 2050 were introduced (Fig. 5.13). The meeting was finished by the explanation of the intention and objectives of the next workshop – the application of the scenario-technique. Therefore, every participant had to do some homework in order to constitute ideas for the creation of new land use patterns for the year 2050.

The aim of the second workshop was to create integrated scenarios as a consensus of the entire group on the basic settings for the year 2050. The group was split up in two parts and manually-drafted versions of the integrated scenarios were presented on wall-papers. The compiling of these interim results took place in the afternoon

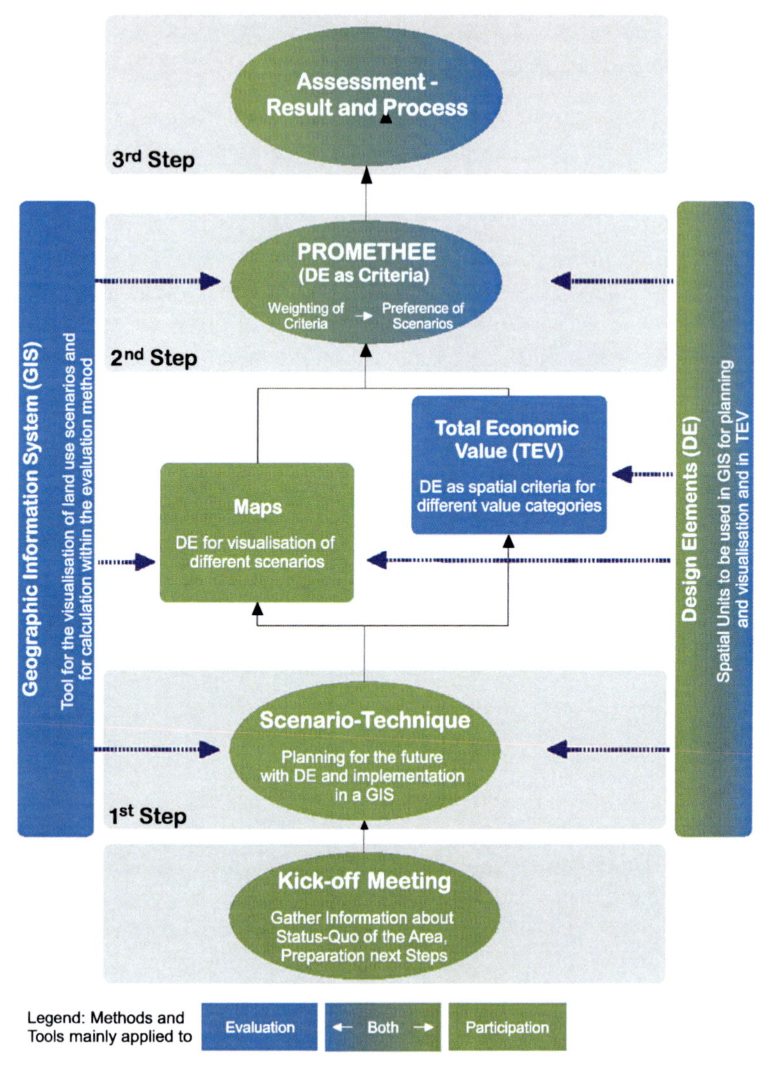

Fig. 5.10 Flow diagram and structure of the participatory integrated assessment process at Nessmersiel

session, where the spokes man of each group presented their results in a plenum. Afterwards, the entire group had to merge these results into one map, so that finally three integrated scenarios "A", "B" and "C" were developed. These integrated scenarios provided the input for the next steps in the process: The implementation of the results into a Geographic Information System (GIS) using the concept of Design Elements (DE's). The manually-drafted land use options were transfered to GIS maps for the application of the outranking method. The different land use options like wind mills, salt marsh, natural camp-site or crop land were implemented via the DE's.

The aim of third workshop was to determine the weights of DE's implemented in the integrated scenarios. The process of determining the weights was based on the *direct-ratio* method and was applied by using a scoring matrix to stimulate cross-sectoral thinking. The purpose of this step was to answer the question: How important is a specific DE? The demand of sustainable development is to integrate social, economic and ecologic aspects. In this process this was achieved by the application of the DE together with the weighting of these DE. The DE are assigned to the different value categories (see Sect. 4.3) and the process of weighting the DE is intended to stimulate multifunctional thinking to achieve cross-sectoral integration. Hence, the final result should show the best integrated scenario fulfilling both the requirement of multifunctionality (balancing different sectors) and preserving ecological functions (introduction of DE of different value categories).

The final meeting of the PIA process was to present the final result of the outranking method and to gain feedback of the participants on their opinions on the entire process and basic questions, e.g. did we meet the expectations and was the process beneficial for the participants? What criticism and what suggestions did the participants have? In such processes, especially if they focus on specific issues burdened by many and long-lasting conflicts, the feeling of participants (i.e. the subjective element) in participation processes is very crucial.

5.2.4 Design Elements, Geographic Information System and Scoring Matrix

5.2.4.1 Design Elements and Geographic Information System

The issue of *multifunctionality* raise the question of tools and methods. From the perspective of spatial planning there are different types of land use and different perspectives which have several requirements on space and resources and also have various effects on nature and on each other. The concept of sustainable development requires taking into account effects on the present land use as well as on possible consequences in the future.

The main characteristics of spatial planning today are the following: it is static and selective instead of dynamic and integrative, more regulative instead of process-oriented and re-active instead of pro-active – see e.g. the recommendations of the NORCAOST project (NORCOAST Project Secretariat 2000). Spatial planning is a more regulatory and sectoral planning instrument which only balances different sectoral requirements, e.g. a road will only be assessed with regard to the benefit for traffic and the impacts on nature and inhabitants, but does not consider further benefits for other user perspectives. This bridge can be built, with the concept of *Design Elements* in combination with participation and evaluation, especially the weighting of criteria.

The implementation of Design Elements (DE) is positioned between the regional and the local level of planning. It is more concrete than regional plans, but not so detailed as local plans (e.g. town planning). The concept of DE is based on the insight, that "coastal resource systems are valuable natural endowments that need to be managed for present and future generations. [...] The intrinsic economic value of coastal resources represents a 'capital' investment for mankind by nature. The goods and services derived from them are the 'interest' generated by the investment. Hence, the destruction of the resource base means depletion of the 'capital' and therefore less interest and the ultimate exhaustion of what nature has freely provided" (Post and Lundin 1996, pp. 3–4).

Consequently, DE are divided into two parts: natural units and (land use) options. The natural units fulfil at least two functions: For example, a salt marsh is of interest to *nature conservation* as well as for *coastal protection*. For example, a tidal creek can provide drainage to the Wadden Sea and a navigation channel for vessels. Applied DE's in Nessmersiel are shown in Fig. 5.11.

The intention of the concept of DE is:

- To provide a joint basis for the discussion and deliberation of the planning steps of coastal management projects, i.e. the participants can use the DE to create a (visionary) land use plan for a certain area,
- to serve as spatial elements for the visualisation and implementation in a GIS, i.e. the results of the planning step can immediately be implemented and visualised,
- to be deployed, equipped with data and information, within a socio-economic-ecologic evaluation process.

The application of the Total Economic Value (TEV) approach using the three value categories *benefit value* and *existence value*, demand an adaptation of the DE to these categories. The implementation of DE within a participation process requires a certain level of detail, so that every participants is able to realise the purpose of a specific DE. For example, the natural units can be differentiated at a plant community level, but in this case it is sufficient to distinguish only between pioneer zone, salt marsh and marsh. The level of detail depends also on the detail available for other DE's. A balance should be given in the level of detail for each DE. The level of detail can be enhanced if necessary for specific purpose. The leading organisation created the first version of the DE's, but within the workshops these DE's were adapted by the participants to local circumstances and e.g. transferred to an applicable scale.

The DE serve as spatial criteria within the evaluation method. Furthermore, the applied DE represent the land use requirements of a certain user perspective for the pilot area Nessmersiel. This does not imply that these DE are used exclusively by this user perspective. This point has been tackled by the scoring matrix, which was generated for the weighting of criteria. This is a new approach to involve stakeholders in the planning step, i.e. in a step where a spatial plan for a certain region is generated. More importantly, the stakeholders themselves have to weight the jointly applied DE's as a group and to find a consensus on future land use options. Within this planning step all relevant user perspectives are involved. Every

Design Element	Benefit/Impact	Type of Benefit	Specification Scenario
Nature Conservation			
Pioneer Zone	Endangerment, regeneration capability,		aggregated functional scoring
Salt marsh	orientation by nature, nature protection status	Information	of...biotope types
Marsh	Value for birds	function/ indirect	...important areas for birds
Conservation Area	Value for nature conservation	use value	...nature conservation areas
Foreshore shape	Value of zoning of vegetation		...zoning of biotope types
Information Centre	visitors/year		Estimated sales per visitor
Salt Marsh: Path	length, information infrastructure	Use value	Estimated costs
Information and Observation	infrastructure for nature information and observation facilities		Estimated costs
Agriculture			
Potential agricultural yield	Potential livestock density, soil conditions	Production function/ indirect use value	Aggregated functional scoring of estimated yield of agronomic and pasture farming
Salt marsh	Livestock density		estimated net yield of biotope types
Cropland	Soil conditions		
Grassland	Livestock density		
Landscape conservation farm	cultivation and research on salt resistant crops, agricultural information path	Use value	estimated increase of net yield comparing to traditional use
Regional added value	salt marsh-cattle, regional commercialisation, renewable primary products		Estimated intensity of regional added value
Recreation/Tourism			
Recreation	Priority areas for recreation, intensity of use, touristic infrastructure, accessibility		aggregated functional scoring ...of estimated capacity, utilization and accessibility of recreational facilities
Disturbance	optical, olfactory, acoustical	Recreational use function/ indirect use value"	...of windmills, biogas power plant, high visitor density
Characteristic	diversity, historical and specific landscape elements		...of diversity of biotope types, and landscape elements
Dune Camp site	Size		estimated proceeds
Sea bridge	Size		increase of attractiveness
Diversity of tourism	windmill with observation deck, reed-labyrinth, farmers market, regional added value	Use value	Estimation of regional added value and proceeds of the different touristic attractions
Energy			
Windmill	Size		Estimated Proceeds
Biogas power plant	Size	Use value	Estimated Proceeds
Photovoltaic cells	Size		Estimated Proceeds
Traffic			
Harbour area	Passengers/year		Estimated Proceeds
Parking place	Users/year	Use value	Estimated Proceeds
Path	Length		Length
Coastal Protection			
Pioneer zone	Erosion control, sedimentation capacity,	Regulation	aggregated functional scoring of
Salt marsh	energy transmission	function/ indirect	...biotope types
Foreland shape	latitude of foreland	use value	...classified foreland latitude
Main dike	Costs		
Summer dike	Costs	Use value	Ratio: costs per useable area
Groynes	Costs		
Second dike line	Costs		
Existence Value			
Existence value	Existence of the Wadden Sea	Non-use value	Measures of Group Valuation

Fig. 5.11 List of design elements applied in Nessmersiel with additional information on the allocation to value categories
Source: Meyerdirks and Ahlhorn (2007a).

party has the chance to influence the planning and to be involved from the earliest stage on.

As mentioned in the previous sections, data and information of adequate quantity and quality is the major challenge in evaluation processes. The TEV approach with different value categories implicates that some categories can not be allocated to units of money, e.g. the existence value (see Sect. 4.3.2). Two main reasons led to the utilisation of verbal categories for certain criteria: the difficulties in adequately determining these units and the agreement of all participants that accurate data and information could not be gathered for all DE's. Since the intention was to rank scenarios the differentiation between options and not the accurate amount of the unit is decisive. Furthermore, the participants agreed on the required level for the detail of criteria. In Fig. 5.11 the specifications of the DE per scenario are given in the last column.

A Geographic Information System (GIS) is an appropriate tool for visualisation and planning. Some types of land use co-exist and some of them are mutually exclusive. In the German part of the ComCoast project the GIS was used for several items:

- To gather, analyse and visualise (spatial) data of different types of land use,
- to visualise and to plan a certain area within the PIA process,
- to calculate necessary figures for the spatial criteria (i.e. DE).

5.2.4.2 Determination of Weights – Adapted Direct Ratio Method

Within the description of the PROMETHEE method (Sect. 4.3.3) there is no advice on how to determine the weights. Weights are information between criteria. They express the relative importance of one criterion against another. Eisenführ and Weber (2003) describe different methods of determining weights within an evaluation process and they emphasise that the determination of weights is the most sensitive part. The basic requirement of determining the weights for the PROMETHEE method within the Participatory Integrated Assessment (PIA) process is to be able to run the procedure with different stakeholders in a certain time frame. That means it should be possible to co-operate on the determination of the weights within the group. Most methods explained by Eisenführ and Weber (2003) are complicated processes that demand for a substantial amount of time. Therefore, for the pilot area Nessmersiel it was decided to apply an adapted *direct ratio* method. A similar procedure to determine the weights was applied by Ruhland (2004) for the decision support of drinking water treatment.

The *direct ratio* method is not highly rated by e.g. Eisenführ and Weber (2003), but it has several advantages for the application in Nessmersiel. The *direct ratio* method rank the importance of one criterion against the others. The importance of each criterion is not essential; what is important is the difference between criteria, which is expressed in figures. The adapted *direct ratio* method designed for Nessmersiel is shown in Fig. 5.12 and was implemented as scoring matrix.

5.2.4.3 Stimulating Multifunctional Thinking

The procedure to determine the weights was as follows:

First, the different user perspectives listed in the top row were weighted. Therefore, the most important user perspective gets "100" points and the others are rated from a range of "10" to "90". The number of points expresses the difference in importance. This importance is only related to the pilot area Nessmersiel and can neither be transferred to other regions nor to the entire coastal zone of Lower Saxony. The existence value is subsumed under the user perspectives, because it

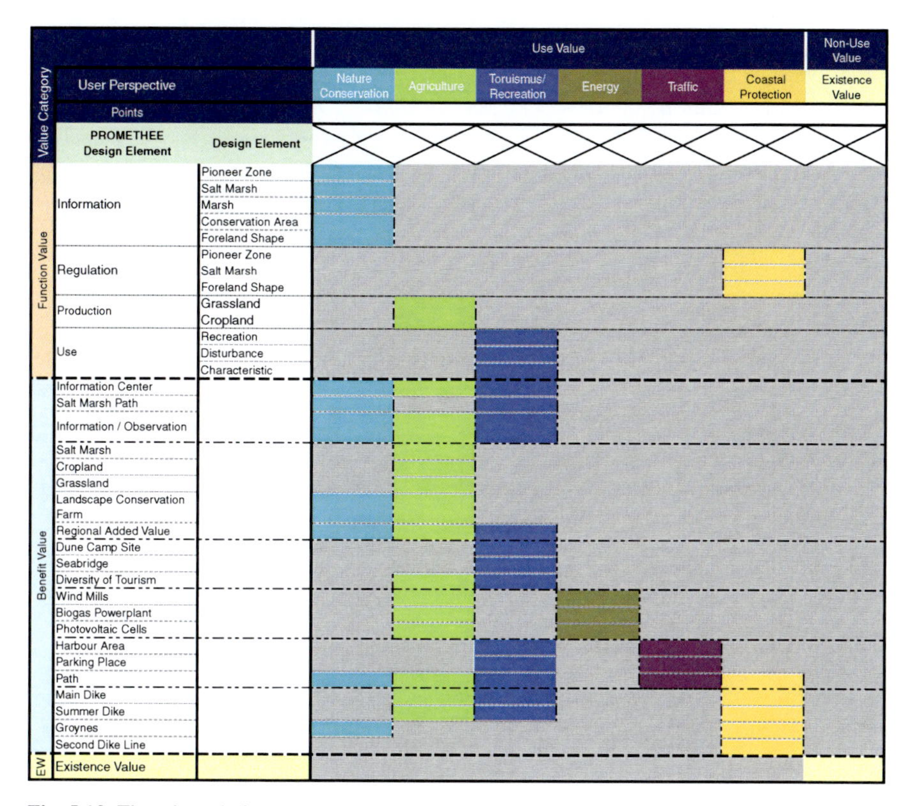

Fig. 5.12 The adapted direct ratio method implemented as a scoring matrix

would have been too complicated to determine the existence value for each criterion. Thus, the existence value was determined for the whole pilot area Nessmersiel (see box on p. 111). After the points given to all the user perspectives were displayed as percentages.

Second, the participants had to follow the same procedure to determine the points for each criterion. The determination of points was done for each column. First, the participants had to define the most important criterion related to the user perspective. For example, which criterion is most important for *nature conservation*? In Nessmersiel, all participants agreed that *nature conservation areas* are the most important criterion. The next step was to give points ranging from "10" to "90" to the other criteria. It is possible to give the same points to more criteria. It is also possible to give "0" points, but that means that this criterion is not relevant. The special effect of this matrix is, that one criterion can get points from different user perspectives. For example, the criterion *Information/Observation* can get points from the user perspective nature conservation, agriculture and tourism/recreation, because this criterion may be important for all of them.

Once the scoring matrix was filled with points, the calculation of the weights followed. The calculation of the weights for each criterion is done by determining the relative importance per column. This step makes it possible to compare the importance of the criteria per user perspective. It is necessary because e.g. "10" points for one criterion under user perspective *nature conservation* can express a different importance than "10" points for the same criterion under the user perspective *energy*. Afterwards, these figures will be added up and the total sum is the weight of this criterion (column). Finally, the matrix shows that the more points a criterion gets, the more important this criterion is, rated in respect of all user perspectives. So, multifunctional thinking is stimulated.

The first row of the completed scoring matrix can be interpreted as the picture the participants have in mind of the pilot area when allocating points to user perspectives. Each participant arranged his points to display how the landscape would be met best, considering different items such as safety and development. Generally, the body of the scoring matrix is the refinement of the decision made in the first row for the user perspective. Now, the participants had to decide which criterion is more relevant than another from the user perspective. Thus, the shape of the landscape of each attendee will be refined by weighting these criteria.

5.2.5 Results of the Process

5.2.5.1 From Basic Settings to Integrated Scenarios

The basic settings for the integrated scenarios were compiled by using information and data from the IPCC for climate change and sea level rise and from national coastal fora (e.g. "Wadden Sea Forum") for input on demographic and socio-economic changes. These basic settings were discussed among the participants and adapted to local circumstances.

In the PIA process a combination of scenario types as discussed in Sect. 4.3.1 were applied. As an input for the first step qualitative and quantitative descriptions of various circumstances like demographic and economic development were introduced. The basic settings like sea level rise, wind speed and precipitation were given as figures compiled from different literature sources. The qualitative part of the scenarios was presented as descriptive, i.e. exploratory, scenarios for the participants. The output of the first step was the consensus on the possible future land use pattern in the pilot area in 2050. The GIS maps showed the anticipated integrated scenarios for the remaining process. Thus, the input of qualitative, quantitative and exploratory scenarios led to the basis for the evaluation process, the GIS maps representing future land use patterns under given circumstances, i.e. the future development of the pilot area as anticipated by the participants of the PIA process.

The basic settings were as follows (Fig. 5.13): *Coastal protection* was widely and highly accepted in the society. The fundamental boundary conditions with regard to climatic change for the year 2050 were set as follows – see e.g. IPCC (2000), CPSL (2001), Evans et al. (2004), WSF (2005), Solomon et al. (2007): 30 cm sea level rise, growth of the fore land and the plates, increase of precipitation up to 10% per year and variation of wind speed of about ±5%.

	Scenario A	Scenario B	Scenario C
General Background	**Successful continuation of recent trends and strategies under good economic conditions**	**Bad economic development and environmental disasters hamper and form the development**	**Successful socio-economic development with a sustainable direction**
Economic Development	• low economic growth • unbalanced distribution of the added value	• in average decreasing growth • unbalanced distribution of the added value	• even economic growth • almost balanced distribution of added value
Demographic Development	• low decrease of population • light superannuating tendency	• low decrease of population • significant superannuating tendency (migration)	• low increase of population • light superannuating tendency
Social and Political Conditions	• unchanged attitude of the Wadden Sea National Park • increasing importance of local and regional decisions on coastal protection and spatial planning (development)	• 2 environmental disasters with negative impacts • National Park has a low social acceptance • reluctant decisions on coastal protection projects and spatial planning (development)	• 2 environmental disasters with positive effects (innovations) • National Park has a high social value • increasing importance of local and regional decisions on coastal protection and spatial planning (development)
Economic Conditions	• increasing energy consumption • high portion of local/regional produced renewable energy • tourism is dominant economic aspect	• increasing energy consumption • low portion of local/regional produced renewable energy • agriculture and tourism relevant economic aspects	• increasing energy consumption • high portion of local/regional produced renewable energy • tourism dominant economic aspect
Regional Development/ Identity	• priority of economic aspects • awareness and conservation of regional Identity	• focused to remove damage • highly focused on regional identity	• sustainability as mission statement • awareness and conservation of regional Identity

Fig. 5.13 Basic scenario-settings for the pilot area Nessmersiel in the year 2050

The morphological boundary conditions, estimated for the fore land in the German pilot area were determined by erosion tendency ranging from 100 to 200 m. This depends on the maintenance of the fore land; the figure is a rough estimation according to today's erosion of approx. up to 4 m per year in some places. Figure 5.14 shows the legend for the maps displayed by Figs. 5.15, 5.16, and 5.17.

5.2.5.2 Result for Integrated Scenario A

The result of the alternative land use option for the integrated scenario A is shown in Fig. 5.15. "The boundary conditions of scenario A can be summarised as a continuation of current trends and business. For the natural fundament the consequence will be that to some extend natural drainage will be applied, the salt marsh will be extensively used by grazing, and the polder will be intensively used by agriculture (crop land). The current *coastal protection* layout will experience some changes: the summer dike will be opened by some breaches. Due to the rising fore land the inundation rates will be the same as today, the protection of the main dike foot will be enhanced as a consequence of the breaches in the summer dike" (Meyerdirks and Ahlhorn 2007a, pp. 10–11). Due to the breaches in the summer dike, the structure and the shape of the summer polder may change in certain places. Erosion of the fore land will take place in the eastern part of the pilot area. To avoid further erosion, brush wood groynes will be installed in the eastern part. The fore land of the western part has a sufficient width, and based on the assumption of accreting fore land with

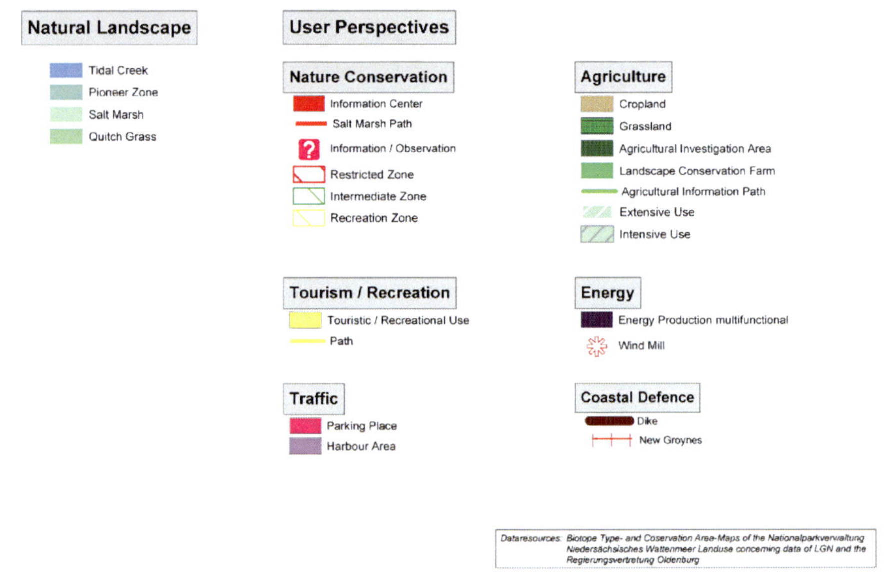

Fig. 5.14 Listing of design elements and legend for the maps – Figs. 5.15, 5.16 and 5.17

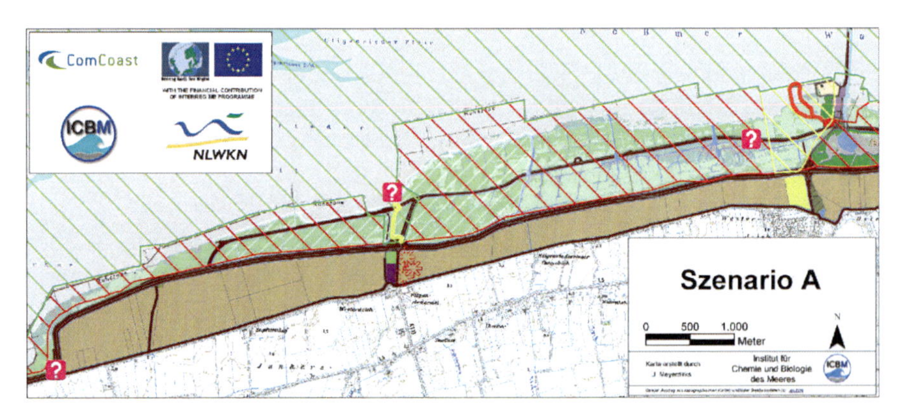

Fig. 5.15 The integrated scenario A

sea level rise, nothing seriously will happen. "*Agriculture* will intensively use the polder, the areas for grazing in the fore land (summer polder and salt marsh) will be more closer to a farm house. A landscape conservation farm will be established within the polder. This farm will mainly use the areas in the summer polder. A local market to sell local products will be established at the farm as well as in the trade or tourist areas near the village Nessmersiel. Natural development of the fore land drainage and enhancement of the information infrastructure will be developed by the *nature conservation* function. The tourist infrastructure will be extended in the proximity of the ferry harbour, mainly temporary infrastructure will be installed. The

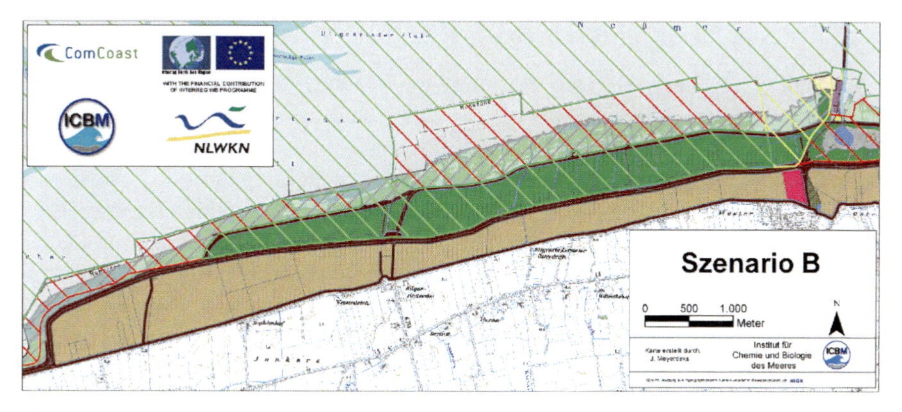

Fig. 5.16 The integrated scenario B

area between the main and the second dike line will be used by tourist infrastructure, e.g. farmers' market. The *energy* function will be characterised by a biomass power plant and cultivation of energy crops. The biomass power plant will be established in the proximity of the landscape conservation farm" (Meyerdirks and Ahlhorn 2007a, pp. 10–11).

5.2.5.3 Result for Integrated Scenario B

The result of the alternative land use option for the integrated scenario B is shown in Fig 5.16. "The boundary conditions of scenario B can be summarised as negative change. For the natural resources the consequences are: natural drainage of the salt marsh, increasing agricultural use of the summer polder and intensive arable use of the polder areas. The *coastal protection* function fails to maintain the drainage of the fore land, maintenance of the summer dike is dependent on costs, but a buffer zone will be created in front of the summer dike to protect the fore land against erosion. The erosion is estimated at approx. 150 m until the year 2050 from now on. The main dike will be heightened if necessary. Some changes will take place due to *nature conservation* measures: parts of the summer polder will be reduced in the level of protection from restricted zone to intermediate zone, and nature conservation will concentrate on the salt marsh in front of the summer dike. *Tourism* and *recreation* will be increased and in consequence the necessary infrastructure will be enhanced ('move to the water'). In the polder area 'fast-food-tourism' will grow (increase in tourists in the harbour area for access to the barrier islands and no consideration of nature and environmental conservation on the part of tourists). Regarding the *energy* function energy plants will be cultivated without local utilisation" (Meyerdirks and Ahlhorn 2007a, pp. 11). In the wake of severe disasters the communication and co-operation between different stakeholders decreases to a minimum. Thus, the orientation of individual land use functions focus more on maximising their own benefits/profits.

5.2.5.4 Result for Integrated Scenario C

The result of the alternative land use option for the integrated scenario C is shown in Fig. 5.17. “Scenario C can be described as the scenario featuring the most comprehensive tendencies towards ecological and sustainable coastal development of the pilot area. For example, a natural drainage system can develop in the salt marsh and in the summer polder. The polder will be intensively used by arable farming. There are several changes in *coastal protection*: the salt marsh will increase in the shelter of the sedimentation fields, the summer dike will be broken down completely, the main dike will be adapted to an overtopping resistant dike and salt water will be restored in the western polder area. The drainage of the main dike will be enhanced and an additional embankment at the main dike foot will be built (as compensation for the destroyed summer dike). A landscape conservation farm will be established as in scenario A. Additionally, the utilisation of the summer polder and the salt marsh will be provided via *nature conservation* measures. As regards *nature conservation* the protected areas will be re-arranged and adapted to the situation following the destruction of the summer dike. The information infrastructure will be enhanced. For *tourism* several changes will take place: touristic utilisation of the pilot area will be increased, but with consideration of the requirements of and in cooperation with *nature conservation*. The main point is that a tourist infrastructure will be established which tries to capture the added value in the area. For example, a promenade will be installed into the Wadden Sea near the ferry harbour and a nature camping site will be included. Main changes will take place for the land use *energy*. A biomass power plant and wind mills will be installed to enable the village of Nessmersiel to be independent from external energy supply” (Meyerdirks and Ahlhorn 2007a, pp. 11–12). Although, in “C” two severe disasters occurred, the willingness to communicate and to co-operate is obvious. The implemented and established Design Elements (DE’s) are of multiple use to different types of land use. For example, in the western part of the polder area a testing field for salt resistant plants will be established to investigate the potential of a new family of

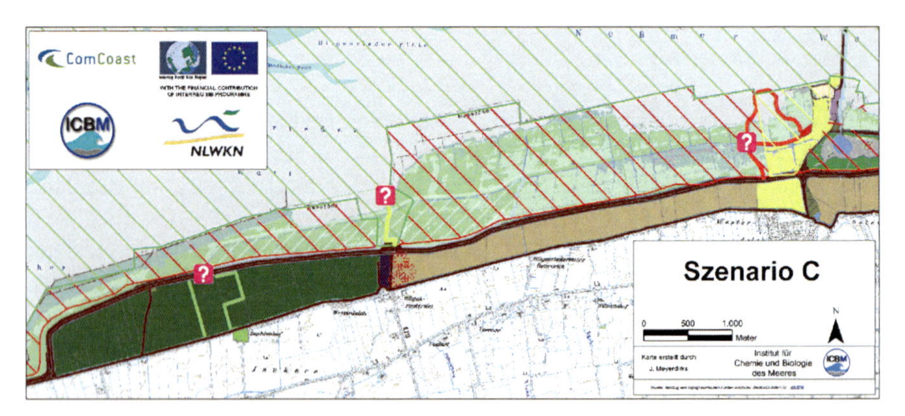

Fig. 5.17 The integrated scenario C

arable plants, i.e. option for diversification of farmers income in combination with new technology to deploy these plants in the biomass power plant. Furthermore, the test field can be accessed for tourists via an information path. The installed wind turbines are equipped with observation platforms for tourists. Local products can be sold at the farmers' market-place in the eastern part of the polder (yellow area).

To conclude, different boundary conditions lead to three different land use options (integrated scenarios) using the same DE's.

5.2.5.5 Outranking of Integrated Scenarios – The Ranking

In Sect. 4.3.3 the theoretical background of the PROMETHEE method was explained. It was stated that PROMETHEE enhances the dominance relation and that the results of PROMETHEE I and PROMETHEE II enrich the decision pocess. In Sect. 4.3.3 technical expressions were used like *out-flow, in-flow* and *net-flow*. These flows indicate the preference for one alternative rather than another. But, these technical expressions are not really catchy for non-experts. For that reason, the technical expressions were renamed: the out-flow was named *agreement*, the in-flow was named *disagreement* and the net-flow was named *extent of agreement*.

The out-flow expresses how an alternative outranks another alternative. Therefore, it can be transformed into *agreement* because the higher the out-flow the higher the preference of this alternative, i.e. the agreement to chose this alternative. The in-flow expresses how an alternative is outranked by the others. Therefore, it can be transformed to *disagreement*. The lower the in-flow the better the alternative, i.e. the higher the in-flow the worser the alternative. That means the disagreement is higher if the in-flow is higher. The net-flow is calculated by the difference of both out-flow and in-flow. The net-flow shows the total agreement to an alternative. The net-flow ranges between -1 and 1. So, that the *extent of agreement* is between -1 and 1, where -1 means complete disagreement and 1 complete agreement.

The calculation of the ranking is exemplified in Fig. 5.18. The multi-criteria matrix is the common basis for all calculations. One column is reserved for one Design Element (DE). Each DE has a different characteristic for each scenario. For example, the criterion *crop land* is related to the value category "benefit value" and can be calculated with real figures in units of money. The criterion *Functional Value Information* was not calculated using money as an unit because it is very difficult to determine the value e.g. for a salt marsh (see Sect. 4.3). Therefore, points are used as to differentiate between the scenarios. The approach on how to evaluate and how to calculate these points for nature conservation is explained comprehensively in Meyerdirks (2008).

The third category of criteria is related to the *information center*. It is difficult to express the value for the *information centre* only in money. Within the scoring matrix the criterion *information centre* was acknowledged as important for the user perspectives *nature conservation, agriculture* and *tourism/recreation*. But, the challenge was to determine what benefit the user perspective *agriculture* might gain in terms of money from the *information centre*. Considering the context of each scenario and with the expertise of the participant from the chamber of agriculture it was

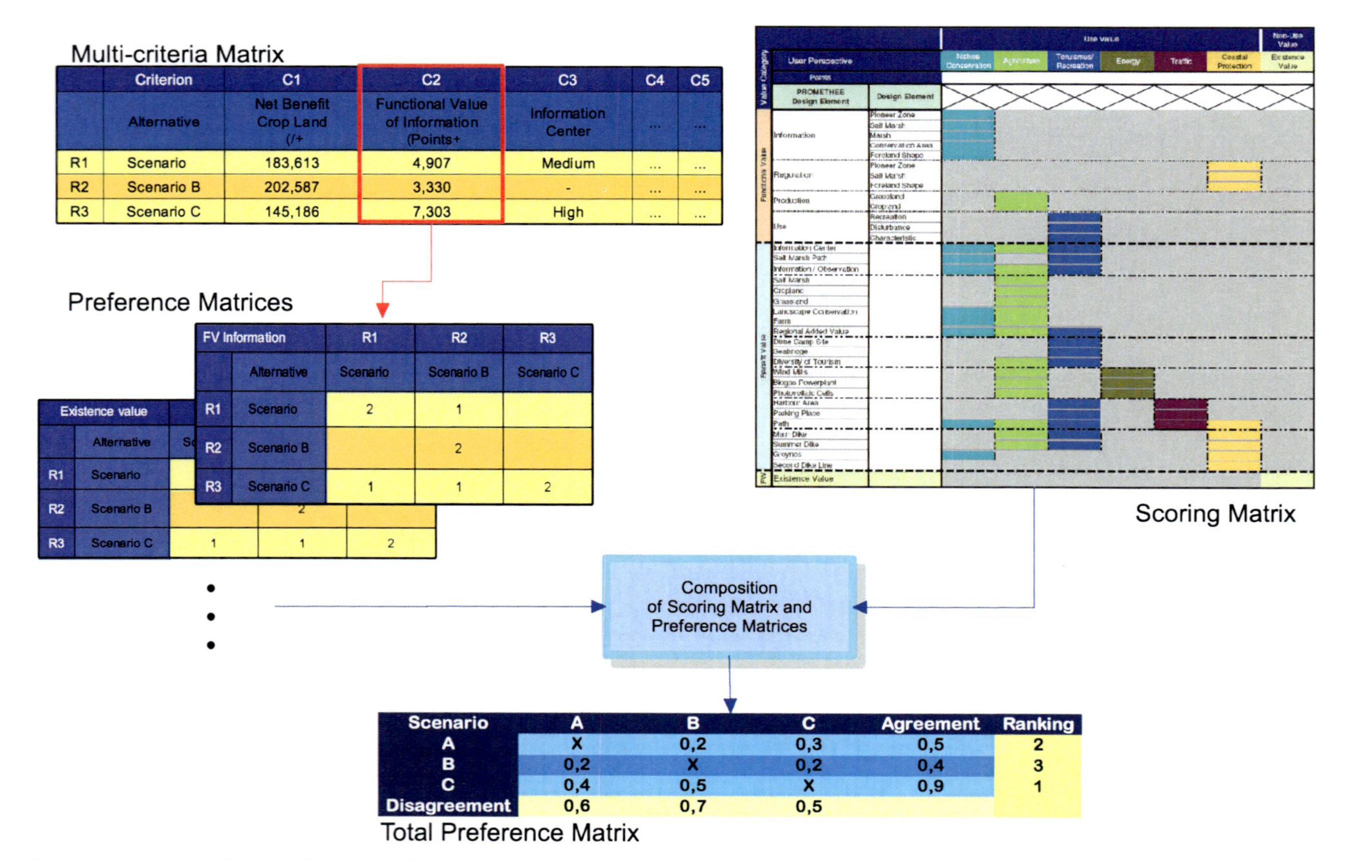

Fig. 5.18 Calculation procedure for PROMETHEE

decided to allocate verbal categories ranging from *low* to *high*. Thus, the criterion *information centre* got a *medium* ranking for scenario "A", nothing for scenario "B" (no information centre installed) and *high* for scenario "C". This classification is the consequence of the enhanced communication and co-operation tendencies implied in the integrated scenario "C". Finally, this was done for all user perspectives which might benefit from the *information centre* and the final characteristic for each scenario in the multi-criteria matrix was determined (see Fig. 5.18).

In PROMETHEE the simplest preference function was applied without any threshold (see Sect. 4.3.3), i.e. without a strong preference. In the case of a difference between one criterion related to scenario "A" and the same criterion related to scenario "B". For example, the criterion *crop land* for scenario "A" is 183,613 € and for scenario "B" 202,587 €, because the area for crop land is bigger and the intention is to maximise the profit per ha. The resulting preference matrix is shown in Fig. 5.18. Equivalent matrices were calculated for each criterion. So, that in the end more than 25 preference matrices were calculated. After the calculation of each preference matrix, each matrix is multiplied by the related weights. The result is shown in Fig. 5.18.

The *agreement* to each scenario is indicated in the right column and the *disagreement* to each scenario in the row at the bottom. Finally, the partial ranking of PROMETHEE I and the complete ranking of PROMETHEE II show the same results: scenario "C" is preferred to scenario "A" and scenario "A" is preferred to scenario "B".

To obtain a detailed insight in the performance of the integrated scenarios with regard to each criterion it is useful to look at the profile of a scenario. The profile of "C" (Fig. 5.19) shows that "C" was preferred for most criteria to the other alternatives (green bars). The integrated scenario "C" is not preferred according to the criterion *crop land*, because in "C" the total area of *crop land* is reduced in comparison with "B". Here, the preference is to maximise the profit, so "B" wins. In "C" parts of the polder between main dike and second dike line will be used as testing fields for salt resistant crops and for an information *path*. The same is valid for *grass land*. In "B" *grass land* will be maintained in front of the main dike as long as the summer dike holds. On the other hand, the functional value *information* is the highest in "C", because of the natural development of the fore land. In "C" the main dike will be adapted as an overtopping resistant dike with a geo-grid under the top layer of clay. The adaptation is necessary because the summer dike will be destroyed and boundary conditions will change e.g. wave climate and the increase of the sea level to about 30 cm. The integrated scenario "B" was also not preferred according to this criterion because the summer dike will not be maintained and may be destroyed by 2050. Consequently, an adaptation and strengthening of the main dike is necessary. In "B" the strengthening of the main dike will be done in a traditional way and that might be much more expensive than the option applied in "C". Finally, "A" was preferred to the others in respect of these criteria. For example, the summer dike in the fore land still exists, albeit with three breaches and an adaptation of the main dike might not be necessary.

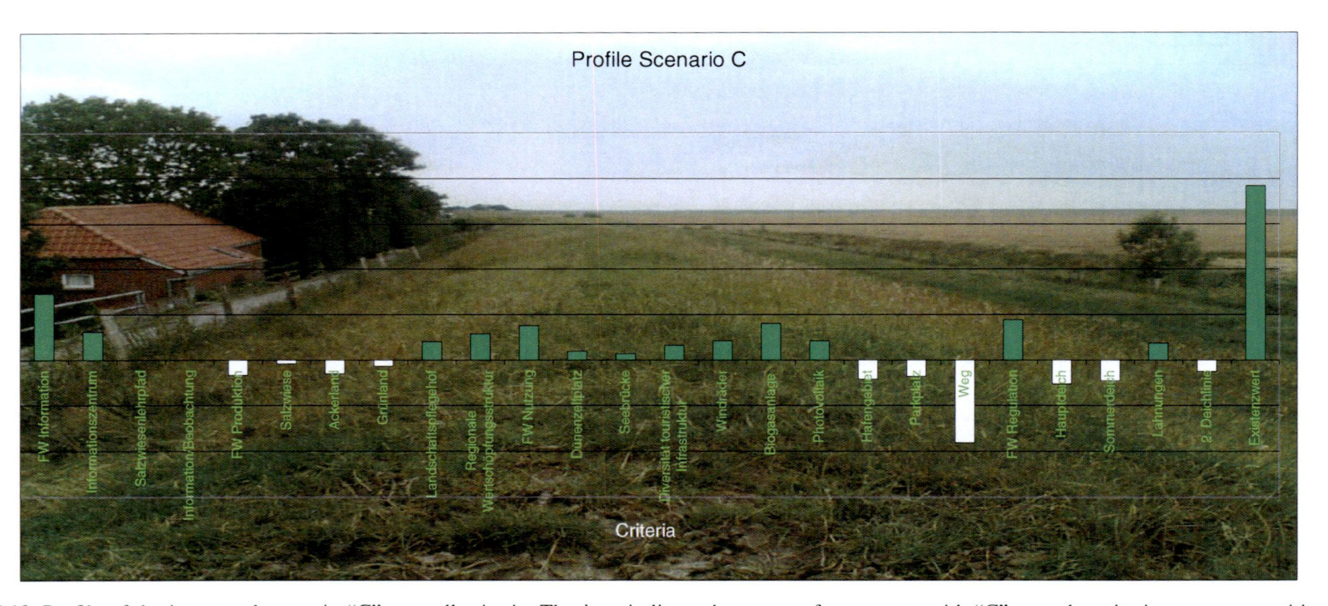

Fig. 5.19 Profile of the integrated scenario "C" over all criteria. The *bars* indicate the extent of agreement with "C" over the criteria: *green* = positive, and *white* = negative

Regarding the second dike line, "B" was preferred to the others, because in "B" no money will be spent for maintenance. In "A" and "C" the second dike line will be maintained and consequently costs will rise. The decision whether a criterion is preferred to another is based on the estimated costs incurred until 2050. These costs are linked to the area of the pilot area, i.e. in "A" the pilot area is about 910 ha, in "B" about 919 ha and in "C" about 999 ha. The assumption is, that the less money spent the more this solution will be preferred, in terms to economic effectiveness. Because of the limitation of the polders adjacent to Nessmersiel these costs can only be referred to the pilot area. Here, the pilot area is treated as a close system. Different approaches have to be considered for different areas.

5.2.5.6 Different Weights – Different World's

In Fig. 5.20 the *disagreement* with the integrated scenario "B" is shown in the results of the one-to-one meetings and the group decision. The bars on the left side from stakeholder 1 to stakeholder 8, represent the different stakeholders and their weightings. The different heights of the bars indicate the different weighting of the criteria and finally the different results. For example, the disagreement with "B" is higher for stakeholder 7 and 8 than for stakeholder 2 and 3. The blue line represents the mean value of the one-to-one results. On the right-hand side of the diagramme the PROMETHEE group decision is shown. An outcome is that for the integrated scenario "B" the mean value of the one-to-one meetings is identical with the group results. This reflects that the group decision-making process was not dominated by a certain stakeholder and that there was a homogeneous negotiation of weights. The results are slightly but not substantially different for the other scenarios. The reason for these differences are exemplified for a selected stakeholder (see box below).

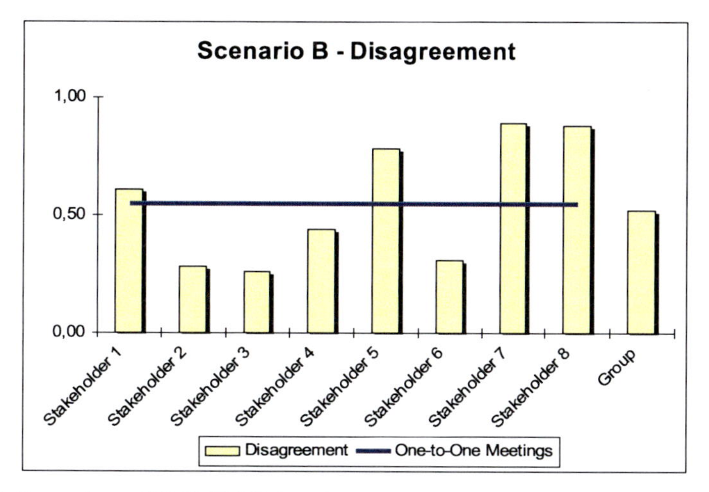

Fig. 5.20 Comparison of PROMETHEE results: one-to-one meetings and group decision

Obviously, these differences lead to the questions: What are the consequences of varying the weights? Do different weights produce to different "world's"? It needs to be borne in mind that the integrated scenarios are results of deliberation and negotiation between all participants during the first consensus workshop. The basis for calculating the values (characteristic per scenario) of the criteria are the integrated scenarios. Thus, the values of the criteria are the basis for determining the preferences of alternatives. These values are fixed for the integrated scenarios of the pilot area Nessmersiel. Consequently, the weights are the parameters, which can be varied to provide insights, onto what scenario will be preferred. This is expressed by weights in the scoring matrix.

Three different assumptions have been simulated: *sectoral weighting, extreme weighting* and *equilibrium weighting*. For *sectoral weighting* no criterion is relevant for more than one user perspective and the existence value is set to "0", because in a sectoral world the shape of the landscape is less relevant. The result is that "C" is preferred to "B" and to "A". In comparison to the previous results "A" and "B" have changed positions. The reason was explained before, for the selected stakeholder (see box on p. 153) (Fig. 5.21).

Extreme weighting is characterised by an existence value = "0", *coastal protection* is prior-ranked, *agriculture* wants to maximise its profits, *nature conservation* is reduced to a minimum and *tourism/recreation* focusses on a fast and high flow-rate of people. The result: "B" is preferred to "C" and to "A". The reason is that "B"

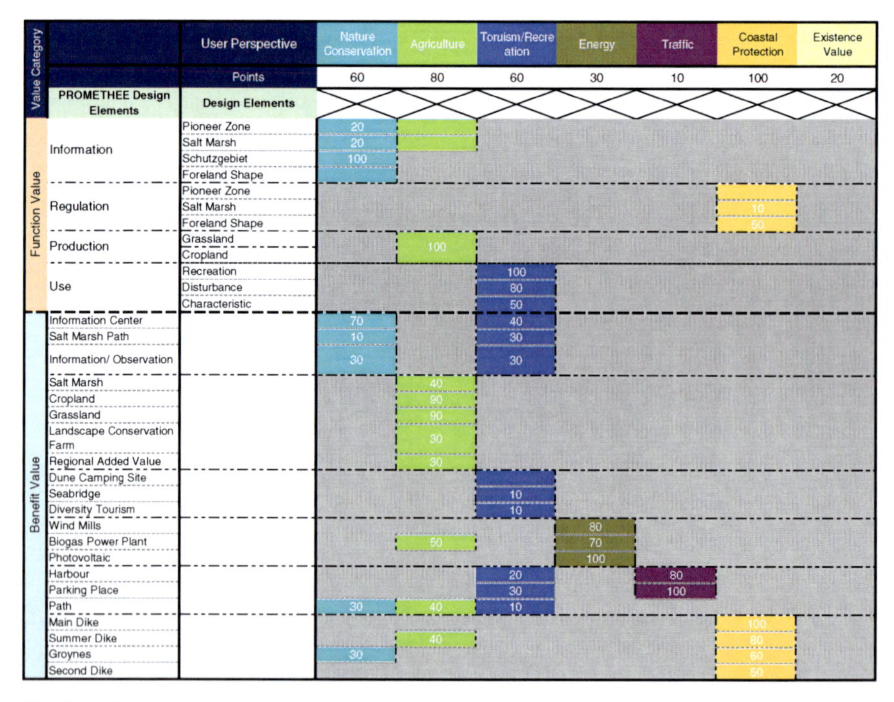

Value Category	PROMETHEE Design Elements	Design Elements	Nature Conservation	Agriculture	Toruism/Recreation	Energy	Traffic	Coastal Protection	Existence Value
		Points	60	80	60	30	10	100	20
Function Value	Information	Pioneer Zone	20						
		Salt Marsh	20						
		Schutzgebiet	100						
		Foreland Shape							
	Regulation	Pioneer Zone							
		Salt Marsh						10	
		Foreland Shape						50	
	Production	Grassland		100					
		Cropland							
	Use	Recreation			100				
		Disturbance			80				
		Characteristic			50				
Benefit Value	Information Center		70		40				
	Salt Marsh Path		10		30				
	Information/ Observation		30		30				
	Salt Marsh			40					
	Cropland			90					
	Grassland			90					
	Landscape Conservation Farm			30					
	Regional Added Value			30					
	Dune Camping Site								
	Seabridge				10				
	Diversity Tourism				10				
	Wind Mills					80			
	Biogas Power Plant			50		70			
	Photovoltaic					100			
	Harbour				20		80		
	Parking Place				30		100		
	Path		30	40	10				
	Main Dike							100	
	Summer Dike			40				80	
	Groynes		30					60	
	Second Dike							50	

Fig. 5.21 Scoring matrix of a selected stakeholder

reflects this "world" best. *Agriculture* is represented in the integrated scenario "B" through the maximisation of production profits. *Coastal protection* is focussed only on the main dike, no money is spent for other elements.

Result for a Selected Stakeholder

The result of the ranking was as follows: The integrated scenario "C" is preferred to the others, but there was no clear ranking in PROMTEHEE I for "A" and "B". The application of PROMETHEE II made clear that "A" is preferred to "B", but with no considerable difference.

The reason is that this stakeholder's weighting shows a preference for sectoral position and hence less multifunctionality was found within the scoring matrix. The scoring matrix of this selected stakeholder shows one special item: The existence value was given the lowest points of all participants. Criteria which are only available in "A" and "C" such as *landscape conservation farm* or *dune camping site* got less or no points. The functional value of *information* and *regulation* gained less points in comparison to other stakeholders or the entire group decision. The missing multifunctionality leads to a decrease of the agreement to "C" and "A" and increases the agreement to "B".

The *equilibrium weighting* leads to the same result as the sectoral weighting: "C", "B" and "A". *Equilibrium weighting* of all criteria reflects the result of the applied preference relation. Weights are not allocated with a certain preference, all criteria are equally preferred. The integrated scenario "C" was preferred to "B and "A", because in "C" DE exist, which do not exist in "B" and "A". For example, the *dune camping site* or *testing fields* with the *information path* in the polder area. So, the ranking of "C", "B" and "A" is obvious. But, what about the difference between "B" and "A"? These scenarios have changed their positions again. Not all DE's exist in both scenarios. But, "B" emphasises the maximisation of profit and if this strong preference relation is applied "B" is preferred to "A" with reference. For example, to the profit from *crop land* and *grass land*. On the other hand, the investment in "B" is less than in "A". In "B" no *information paths* are implemented. For these paths it is difficult to determine the benefit, so it was decided to incorporate only the costs of information paths. Consequently, if there are no paths in "B" than "B" is preferred to "A".

5.2.6 Single and Group Decision-Making Process

5.2.6.1 Single Decision-Making Process

The outcome of the workshop was difficult to predict, because there were many fields in the scoring matrix, which exhibited a difference of more than 70 points between each stakeholder. These differences might serve as an indication for possible conflicts in achieving a group consensus (Fig 5.22). The red framed boxes

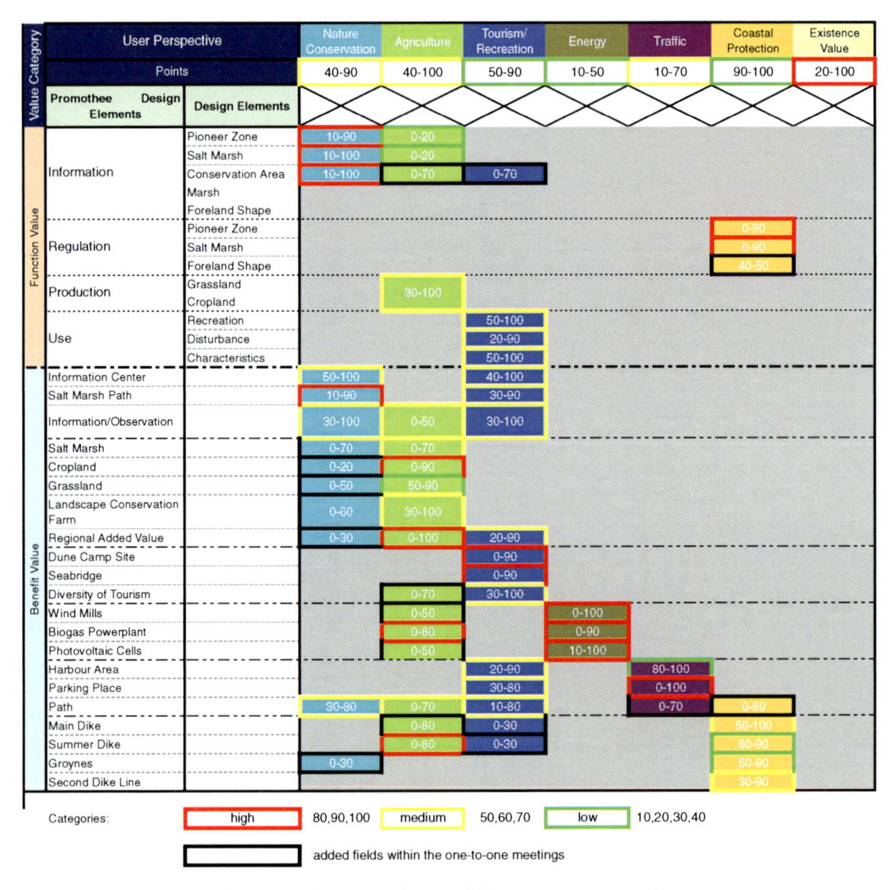

Fig. 5.22 Scoring matrix indicating the variance of the one-to-one meetings

indicate the high differences between stakeholders (ranging from "80" to "100" points), the yellow framed boxes indicate medium differences between stakeholders (ranging from "50" to "70" points) and the green framed boxes indicate low differences between stakeholders (ranging from "10" to "40" points). The black framed boxes were added by stakeholders during the one-to-one meetings. Those were not included in the variances because not every stakeholder added the same box to the scoring matrix.

In the first row only the "existence value" has a red framed box and the points vary from "20" to "100". This difference might originate in the complexity of this value category. All other user perspectives show differences which are medium to low. For *coastal protection, tourism* and *energy* they are low. This might due to different reasons: every stakeholder stated that coastal protection is and was a necessary precondition for living and working in that area. *Energy* was seen as a lower-ranking user perspective in the pilot area Nessmersiel. *Tourism/Recreation* shows also no great discrepancies, because participants acknowledge the significance of *tourism/recreation* in this area. *Nature conservation, agriculture* and *traffic* show a bigger difference. For example, the user perspective *traffic* is not very important for that area, because the pilot area consists mainly of the polder areas between the dikes with adjacent infrastructure like the ferry harbour and a car park. The traffic infrastructure in the pilot area will not be enhanced considerably.

The main body of the scoring matrix has several fields which show higher differences. Here, they should not all be discussed in detail, but some will be chosen to illustrate the differences.

First, the differences strongly depend on the amount of knowledge each participants gained about the other user perspective. In the one-to-one meetings, stakeholders tried to guess which criterion might be the most important for each user perspective. This explains for example the differences on the *nature conservation* criteria. For example, the big difference for the criterion *summer dike* under the user perspective *agriculture* reflects the picture of the landscape the participant might have. Some attendees did not want a *summer dike* in front of the main dike, because of the interruption of the natural processes for salt marsh development. Others wanted to have the *summer dike*, because of the utilisation of the area and the expected advantages. On the other hand, differences may result in the problem of deciding whether this criterion is important or not, e.g. for *agriculture* a *biomass power plant*. A few red framed boxes can be found for *tourism/recreation* and *traffic*. The user perspective *energy* has only red framed boxes; here the stakeholders completely differ in their judgments about the relative importance of the criteria. The arguments against *wind turbines* are as follows: in the proximity of the pilot area many wind turbines had already been installed, so there was no necessity to install additional wind turbines. If renewable energy were to be installed in the area, most stakeholder prefer *photo voltaic cells* on houses. The relative importance of the criterion *biomass power plant* was due to the interlinkage to *agriculture*. The points given for *coastal protection* only show red framed boxes for the criteria relating to the functional value (see Sect. 4.3.2). Most attendees scored these criteria relatively high because they thought that also the *pioneer zone* and *salt marshes* are important for *coastal protection*. Others gave these criteria a low score, because they feared that other criteria considered for the user perspective *coastal protection* did not get enough points and therefore were not given the desired weight. Almost all participants allocated "100" points to the *main dike*.

5.2.6.2 Group Decision Making Process

For the procedure of the group decision process, the group was divided into two parts. Within these two smaller groups the scoring matrix was deliberated and negotiated, because smaller groups provide more advantages of direct discussion between participants. The procedure for determining the weights was agreed within the group (result see box). This went well, but the time scheduled for the deliberation and negotiation phase within these smaller groups was too short. After finishing the smaller group phase, the original idea was to come together and to discuss the smaller group results within the whole group. The same procedure was performed as in the consensus workshop on the planning exercise. These groups did not finish their work in time and hence the unsettled user perspectives were discussed in the plenary session.

Procedure to Allocate Points Within a Group

First, a ranking of all user perspectives or criteria was made. Second, the most important user perspective or criterion was given "100" points. Third, the other user perspectives or criteria were listed in a descending order. Fourth, the difference between the user perspective or criteria were determined. Fifth, points were allocated to each user perspective or criteria. [In the case that experts tried to dominate the decision of the group, first let the non-experts judge on a specific criterion and afterwards the experts. In the consensus workshop there was no need to apply this rule, the participants agreed upon a common strategy on allocating points.]

Taking the results of the single decision into account, the red framed boxes may indicate the conflicts within the scoring matrix. In fact, it was a hard discussion about the points for the *existence value* in the workshop. The problem was to weight between *coastal protection* and the *existence value*. However, *coastal protection* is an important user perspective, but the *existence value*, as defined, encompassed the entire landscape and their shaping ingredients like natural and cultural units. Some participants argued that *coastal protection* was the most important user perspective in the pilot area. But, on the other hand, some participants argued that the *existence value* indicated the fundamental characteristic of the landscape as defined in the questionnaire (see box on p. 111). The *existence value* is difficult to describe, but it reflects which picture of the landscape should be conserved. Because the participants did not agree on the points, they decided to vote. The result of the voting was, that the *existence value* got "100" points and *coastal protection* "90" (Fig. 5.23).

One main modification of the scoring matrix was made by the group, namely the criteria for the functional value of nature (i.e. mudflat without vegetation, upper salt marsh, etc.) were merged to three criteria: pioneer zone, salt marsh and marsh. The group decided that the proposed classification was not adequate and was not reflected by other criteria.

Mainly, the criteria for *nature conservation* gained points in the upper part of the variance interval. For example, the criterion *landscape conservation farm* got a higher score than in the single decision process. The criterion *groynes* got "70" points, i.e. more than 40 points higher than in the single decision process. The group decided that brush wood groynes would have a positive effect on the fore land. For the user perspective *agriculture* no significant modification was made. The points for the criteria show no big differences. Only the criterion *information centre* was added to the list and got "40" points. For *tourism/recreation* all points given for the criteria were in the upper part of the interval as shown in Fig. 5.22.

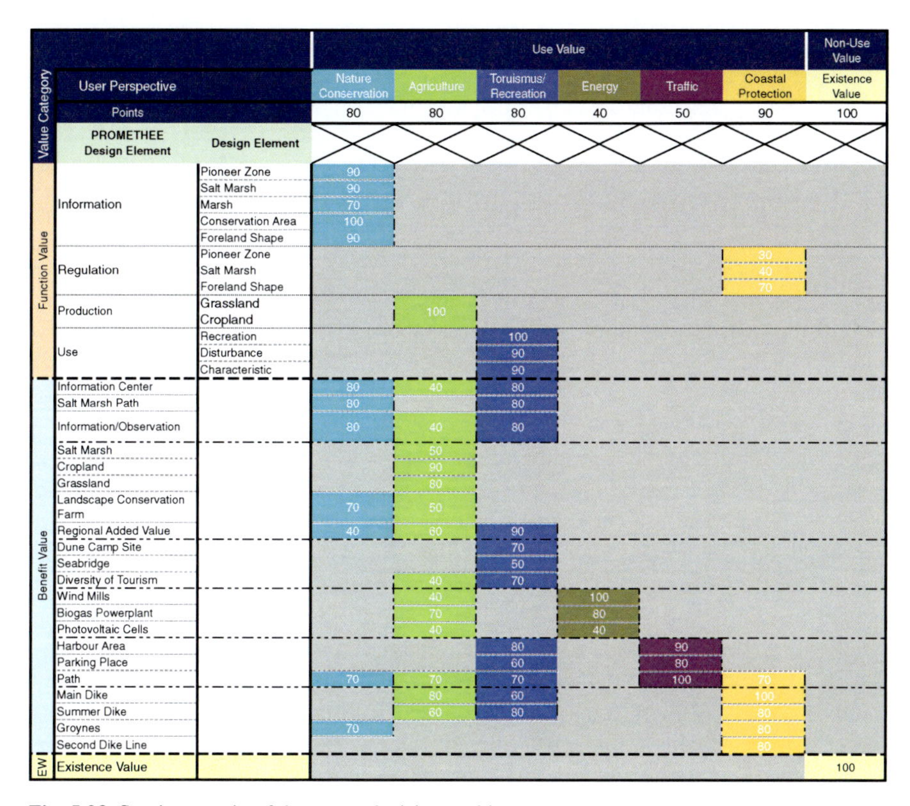

Fig. 5.23 Scoring matrix of the group decision-making process

The group discussed the importance of the user perspective *energy* for the area and concluded that *photo voltaic* was not so important, because of the small number of houses. *Wind turbines* in combination with other features like an *observation platform* for tourists might raise the acceptance, contrary to the single decisions. The *biomass power plant* got a higher score because of the interlinkage to *agriculture*. The criteria *harbour, parking place* and *path* were given slightly the same scoring, because they were seen as important for the area. The *path* got "100" points because of the accessibility benefit to tourists. The resulting points for *coastal protection* indicate the above mentioned effect of taking into account the strategy of giving the points at the right place in the scoring matrix. The allocation of points to other DE for *coastal protection*, did not have a strong influence on the most important criterion, the *main dike*. The criteria *pioneer zone* and *salt marsh* scored very low against the single decision process. This result may indicate that the fore land is not very important for the user perspective *coastal protection*, even if this contradicts some publications.

Consequently, this shows both the advantage and the disadvantage of this method to determine the weights. The disadvantage is that stakeholders can try to influence the arrangement of points in favour of a particular criterion, but this is valid for the most methods. But, within the scoring of the criteria *all* participants can discuss the number of allocated points, and thus can directly influence the result.

5.3 Lessons Learned – Participatory Integrated Assessment

First of all, the Participatory Integrated Assessment (PIA) process at the pilot area Nessmersiel was the first ever conducted at the Lower Saxonian coast. The sufficient state of the coastal protection system led to the decision to run a PIA process to initiate a sustainable coastal development process. In Nessmersiel 2050 was selected as a time frame. The description in the previous sections shows that the process ran well in most of its stages.

Lessons learned at the beginning of the process were that the preparation of the participants at each step of the entire process was crucial. The participants should be very well prepared, before such a process starts. In the case of Nessmersiel, it was decided to invite the potential participants individually by phone. Within the kick-off meeting the aim and the intention of the process and the ComCoast project were explained. There was little feedback after this meeting and few responses to the invitation of the first workshop, which highlighted a problem (i.e. misunderstanding) in communicating objectives and intentions. This problem was resolved by one-to-one meetings with certain participants. Here, the participants expressed their concerns and problems with the project and the process. These meetings offered the opportunity to discuss problems and concerns in more detail than in the kick-off meeting. Nevertheless, the first feedback of the participants was that they appreciated their early involvement in such a process. The reasons given were all similar: the opportunity to discuss issues of a certain area with different representatives of other user perspectives. The participants appreciated the chance of learning more about the other user perspectives.

The exercise of generating three land use options for the integrated scenarios depended strongly on the ideas and the willingness of the participants. New ideas were born in the workshop and some of them were implemented in the scenarios. Even solutions which were formerly not favoured such as the opening or even destruction of the summer dike, were implemented in "C". Here, the PIA process benefited from previous participation processes conducted in the area of the dike board Norden (NLWK Norden 2003) or within the project group for the enhancement of the coastal protection procedure management (NLÖ 2000). These processes served as a good starting point for the PIA process. One of the problems in group decision processes is always to go beyond the status quo of an area. Another important point is to have the right people in the group. The leading organisation should strive to have different kinds of people within the group. For a treatment of this potential barrier see e.g. Houtekamer et al. (2007). Roose (2006) made an investigation on innovation in strategical planning processes for the province Schouwen-Duiveland in The Netherlands and recommended to overcome this barrier by starting a top-down approach initiated by the government. The experience from the Nessmersiel case was that a process conducted by a neutral party can offer advantages, because participants of different organisations and institutions do not feel patronised. The advantage of such a process is, that each participating party has the same status. The same experience was made within the pilot area Perkpolder (NL) in the province Zeeland. The local municipality wanted to improve a former harbour area and initiated a process with different stakeholders (see Appendix B.3). The same positive experience of social learning and active stakeholder involvement has been made in a participatory process for different land use options in the northern Mediterranean (Patel et al. 2007).

For group decision processes Eisenführ and Weber (2003) pointed out three main requirements for the composition of the group:

- cohesion of the group,
- power and status differences,
- individual interests.

For the PIA process these requirements were taken into account. Lack of cohesion in the group can lead to fatal errors in thinking, because the consequences of decisions are underestimated (Eisenführ and Weber 2003). The intention of the PIA process was to support a sustainable coastal development approach and the necessary decision-making process in a certain area by 2050. This was executed by three integrated scenarios. If real planning were to be conducted, the alternatives and the participants would have to be extended. The power and status differences were not important in the PIA process because it was a fictitious planning process, so no decision was taken. But, the land use options were generated as real options. In making their decisions, all participants reflected on the possibilities and available opportunities. This does not imply, that the land use options for the integrated scenarios may not differ in a real planning exercise. The biggest difference is that here a long-term vision was created and not a short-term plan. Furthermore, several items have to be taken into account if a "real world" experiment is conducted – see e.g. Royal Haskoning (2007). It is difficult to investigate the individual interests of each participant in detail within this process. But, as explained in the previous section on the scoring matrix, the interests of certain institutions can be identified via background information justifying the allocation of points to a certain criterion. Again, individuals strive to maximise their benefit. The individual (institutional) interest was investigated within the one-to-one meetings. In these meetings the scoring matrix had to be filled out, so that a comparison could be done between the single and the group results (see previous section). The assumption in bargaining theory is that groups do not have a common goal (see Eisenführ and Weber 2003), this was supported by the differences in single and group decisions on the scoring matrix. But, this did not hamper the discussion in the consensus workshop. To avoid the overloading of participants within the last workshop, where several new items such as the evaluation method and the procedure of filling in the scoring matrix had to be understood, it was decided to run one-to-one meetings before the workshop. This procedure was confirmed by the progress of the workshop itself.

Unbiased interviews were held with the attendees of the PIA process by two students, who investigated the opportunities and challenges of spatial coastal protection concepts in Lower Saxony – see Flügel and Dziatzko (2008). The interviews were intended to give insights in the practical work and cooperation between institutions and organisations dealing with issues at the coast. The representative from the Dike Board stated that initial scepticism was dispelled after a one-to-one meeting, where the concerns were openly and comprehensively discussed. And the experience of being part of such a participation process was rated positively. The representative from the Chamber of Agriculture stated that the process was well-managed and that he had learned from this participation process, but when it came to real planning, more stakeholders would have to be integrated in the planning exercise, especially the landowners in the area of concern. The interviewee from the NLWKN stated that the process reduce the safety of the area, because in one of the integrated scenarios the summer dike would be destroyed. On the other hand, cooperation in compiling plans and programmes using a participation process would take too long time. The interviewee from the National Park management had a contrary impression of the participation process and appreciated the approach to jointly develop plans for coastal areas. If a plan, for example the Master Plan for Coastal Protection, were to be compiled by the responsible institution, less opportunities would be available to discuss alternative options for a problem. In case of the Ley Bay, this led to legal action and the original plans to close the Ley Bay had to be revised. The result of the process was that the Ley Bay was not closed, and the unique lagoon with its salt marshes and natural drainage system was preserved. Sometimes such decisions may lead to expensive solutions, but if the process and the project is planned jointly an additional benefit for others may be gained. The Ley Bay plan was an outstanding example for the approach of "decide – announce – defend", but this should and can be avoided in future (see box).

In the Netherlands the approach to tackle the Weak Spots of their coastal protection system is embedded in a wider approach – see MVenW (2002). These Weak Spots are classified according to the spatial quality of the area (applying seven criteria). Applied solutions should comply with sustainable coastal development as defined for the Dutch coastal zone.

Ley Bay – Part II

In the 1960s a plan was developed to enclose the Ley Bay, because of sedimentation problems: limited access of ships to the harbors and limitation of the drainage of the hinterland. Furthermore, the Master Plan of 1973 stated that the height of the main dike was not sufficient (MELF 1973). These enclosure plans were also mentioned in regional planning programmes. The first idea for the enclosure of the Ley Bay took into account different types of land use, such as agriculture, recreation and nature conservation. But, the survey only took the development of the enclosed bay into account, i.e. after a certain time the ecosystem would have changed from a saline to a freshwater environment.

This would have caused a tremendous change of the flora and fauna of this ecosystem. In the late 1970s the interests of nature conservation, especially for the conservation of the Wadden Sea habitats, attracted increased attention.

The establishment of these interests demanded new solutions to the Ley Bay problems, like water management and shipping. In 1980 the government of Lower Saxony cancelled the planning approval of a complete enclosure and recommended to maintain the natural status of the Ley Bay as saline environment. Afterwards, a group consisting of engineers as well as representatives from nature conservation bodies agreed on a compromise for the required work in the Ley Bay, the "Ley nose". The issue was taken to the European Court of Justice in Luxemburg which decided that only outstanding reasons can lead to the reduction of conservation areas, whereas economic aspects are irrelevant. This decision has had large-scale consequences for further (coastal protection) projects in the entire European Union. Today, the Ley Bay project is completed and the Ley nose has already been built (Fig. 5.24) – Janssen (1992). The impact of the construction on nature is documented by a long-term monitoring process – e.g. Arens (2000), Götting et al. (2002).

Fig. 5.24 Aerial photograph of the Ley Bay
Source: NLWKN (2007a, b) www.nlwkn.de.

To conclude, the Participatory Integrated Assessment (PIA) process is a contribution to the ICZM recommendations of the EU (EC 2002), which emphasise the integration of different stakeholders and a comprehensive approach to solving a problem. The PIA process was initiated to test such a procedure in the coastal zone of Lower Saxony to implement MCPZ. The design of the process with workshops, one-to-one meetings and participants from different fields reflected the ICZM aspects. Especially, the evaluation method PROMETHEE with the scoring matrix to determine the weights of criteria, stimulates stakeholder to discuss proposals from a variety of perspectives.

5.3.1 The Scoring Matrix – Catalyst for a Desired World?

As stated before, the outcomes of the single scoring matrices provide insight into the relevance of Design Elements (DE's) and into the latent conflicts which might occur during the group decision phase. The latter has already been discussed. The variation of the weights lead to different worlds. The case study demonstrated exemplarily: If the weights were only allocated from a sectoral perspective, the integrated scenario "B" would be preferred to the others. But, the integrated scenario "C" is ranked higher than others, if multiple relevance is given for a single DE or all weights are equally allocated. The tenor of "C" is that cooperation and communication functions well amongst all relevant sectors, despite the fact that an environmental catastrophe took place. This led to various and to some extent innovative ideas in the pilot area: Wind turbines may be installed with an observation platform for tourists; in the polder area a testing field for salt resistant crops may be established with the opportunity to educate and to inform interested people; the temporary installation of a dune camp site in front of a main dike, although the area is highly protected; the summer dike may be destroyed to allow tidal influence and the restoration of the summer polder to a salt marsh.

Taken together, all these elements are the expression of a jointly developed and planned area for the future under consideration of the basic settings given in the first workshop (anticipating the future). Taking the objectives of sustainable development into account, this might serve as an example for sustainable coastal development. Three integrated scenarios have been developed as a reaction to three different boundary conditions: The integrated scenario with the highest degree of multifunctionality was ranked highest. This was the final result of the group decision phase, although single decision steps show different results. Thus, the deliberation and negotiation of the relevance of DE's is a major step in achieving a sustainable coastal development. Because all participants had the opportunity to influence the planning procedure various aspects, represented here by different attendees, could be considered.

This visionary planning does not pretend to be an optimum, and should be revised and adapted if new demands or interests emerge. So, a further advantage of this process and its components is its adaptability to new circumstances.

5.3.2 A Vision for the Future – What About the Adaptability?

The result of the outranking method surprised some participants. They expected “A” to emerge as the winner (see voting for the existence value in Sect. 4.3.2). Nevertheless, the entire participation process stimulated the participants to think about future interdependencies. If the integrated scenario “C” was set as a vision for the year 2050, for which all stakeholders strove for, what about changing circumstances and new demands in the future? The participants asked for an iterative process to integrate future demands in the vision. Take for example the *coastal protection* need to heighten the main dike in the proximity of the pilot area in the next 30 years. Construction material (clay, sand) is needed: The question is, where to gain this material. As already mentioned, the established National Park led to conflicts about the utilisation of salt marshes, one aspect affects the excavation of clay in the fore land. Since excavation is prohibited in the National Park, the material has to be excavated in the hinterland. The polders between the main and the second dike line may serve as excavation area. Consequently, there might be a conflict arising on soil between *agriculture* and *coastal protection*.

The conclusion of the discussion on the adaptability was, that an iterative process, i.e. a second loop, is able to adapt the current vision to a new demand (Fig. 5.25).

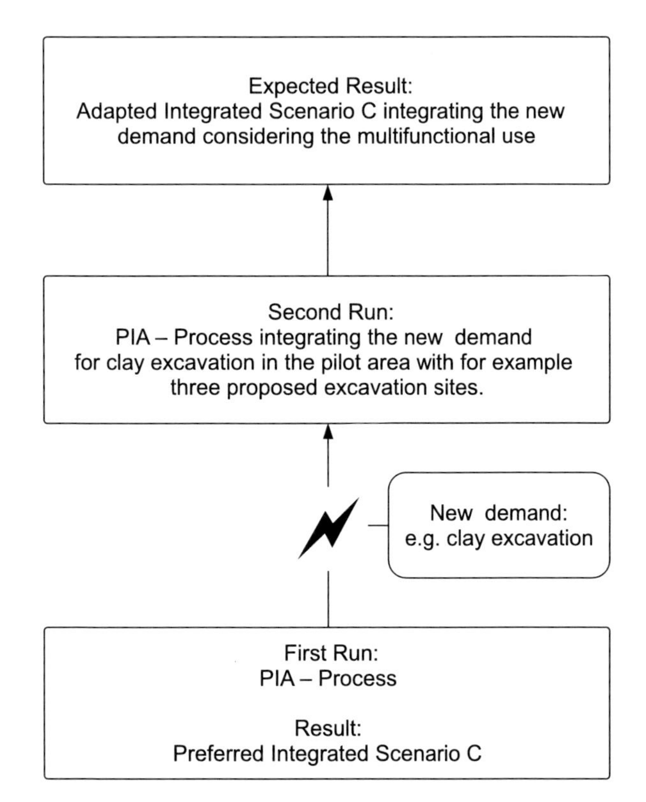

Fig. 5.25 Iterative participatory integrated assessment (PIA) process to integrate new demands

This assumes that the final result of the first Participatory Integrated Assessment (PIA) process is set as a vision and serves as an input for the second loop. The process has to be conducted in a similar way, especially with regard to the weighting and the procedure on how to integrate the whole (excavated material) in a multifunctional (and sustainable) way.

Furthermore, it is possible to revise the results of the first PIA process not only for a new demand, i.e. a single project, which would be necessary, but also if the opinions and the global or local boundary conditions change. A second loop will offer the opportunity to re-think the preferences and the weights of DE's already implemented and to add new DE's if necessary.

Chapter 6
Looking Ahead – A Sustainable Process

Contents

The preamble of the United Nations – Agenda 21 document states (WCED 1987, p. 110): "[...] The broadest public participation and the active involvement of the non-governmental organizations and other groups should also be encouraged. [...] (preamble 1.3)". The transformation of the industrialised society to the knowledge society broadens the perspective from solely economic to include economic, ecologic and social aspects. This was expressed by the Earth Summit conference in 1992. The demand to consider the ecologic, economic and social aspects in further development requires a comprehensive knowledge base. But, this demand makes the world even more complex than before, and the requirements to be met by science and the political system are even higher. Munton (2003) summarises the uncertainty about the environmental and sustainable context as follows: "Indeed, in the absence of overwhelming scientific evidence, in a plural society there will be a plethora of 'right' answers. This suggests the need for inclusive processes that encourage debate, learning, adaptation and consent."

The continuous percolation of additional interests and needs on top of existing approaches calls for adaptation. New adaptation strategies are also required for the likely changes within the natural, environmental and in the social system. Until the 1960s the intention of coastal protection in the low-lying areas of the southern North Sea was twofold: to protect people and property against flooding and to enhance arable land for agricultural use. The upcoming interests of nature and environmental conservation led (or to some extent forced) reflection on the existing approaches. This took place over the last ten decades and will proceed in future. Therefore, the adaptation of existing laws and instruments is required, as well as the provision of appropriate methods and processes. The overall aim after Rio should be sustainable development, and with a coastal focus, sustainable coastal development.

F. Ahlhorn, *Long-Term Perspective in Coastal Zone Development*,

DOI 10.1007/978-3-642-01774-2_6,

The outcomes and experiences of the Participatory Integrated Assessment (PIA) process in Nessmersiel gave evidence of achievements in promoting sustainable coastal development. This PIA process was successful. Although the basis for the process was a fictitious task. The extent of the success should not only be measured by the outcome of the process, other aspects document its success:

- After initial scepticism on the part of different participants, all participants stayed with the process until the end.
- Many participants appreciated their early involvement to the process.
- Although, the results were surprising, these were acknowledged and accepted by all participants.
- All participants appreciated the cooperation and communication elements of this planning exercise.

6.1 Progress Through Adaptation

The answer to the question: "Are the existing strategies equipped to meet the new challenges imposed by climate change?" would appear to be simple. Since the establishment of the Intergovernmental Panel on Climate Change (IPCC), the periodically published assessment reports have given evidence of climatic changes influenced by mankind. These changes are not limited to the biosphere, all parts of the social network will be affected. Prominent aspects of the Global Warming issue are the rising sea level and the concentration of CO_2 in the atmosphere. Obviously, the global effects will have local impacts. Recently, research programmes were launched to investigate the impacts of global climate change at the regional and the local level, down-scaling to determine the impacts to a certain region – see e.g. Jonas et al. (2005), Spekat et al. (2007), Grossmann et al. (2007), Von Storch and Weisse (2008). On the other hand, the options for adapting and mitigating the effects of climate change will be investigated, in order to prepare society for the future e.g. Zebisch et al. (2005), EEA (2005, 2006). Also, the consequences of globalisation lead to a more complex world. Local action can have global consequences, and vice versa. The world is a plethora of different interests and needs, which have to be balanced to achieve co-existence.

At a local level people settled in the low-lying area of the southern North Sea and have to cope with the threats of the sea. The strategy was to adapt to the changing circumstances, because people were only able to react to higher storm surges. Over time the capability of protecting themselves against storm surges grew, but they were not able to predict the height of the next storm surge. The embankments, mainly dikes, were built on intuition and experience. Within the last two centuries a single defence line was built to protect people and their property against flooding. The concept was to have one line of defence (as straight/short as possible) against the sea. This strategy was questioned and modified by the introduction of the different needs of nature and environmental conservation. Parallel to the global change in attitudes towards a sustainable development, in Lower Saxony the first nature lobbyists appeared. One result, for example, was an adapted plan for the coastal protection project *Closure of the Ley Bay*. Consequently, coastal protection has to consider nature and environmental aspects in future projects. This has been supported by laws, ordinances and rules for environmental and nature conservation. Nowadays, further interests and needs, and especially the demand for sustainable development impose new challenges to coastal protection. The demand to adapt adequately to the likely consequences of climate change, will lead to a re-consideration of the current strategy. Currently, the answer to the question whether the existing strategy has to be revised is "no", but this may change over time. The argument quoted is usually that decisions will only be taken on the basis of secure knowledge (Lange et al. 2007). But, efforts in research for coastal protection are focussed on two concerns: to understand and to investigate the failure mechanism systems of embankments and to investigate the probabilistic risk approach to design coastal flood defences – see e.g. TAW (2000), Oumeraci and Kortenhaus (2002), Kortenhaus (2003), Mai (2004), RIKZ (2006), Kortenhaus et al. (2007), D'Eliso (2007), Richwien and Niemeyer (2007). The insight that the days of the existing concept of continuously heightening the dikes are numbered is evident. An assessment of the existing problems e.g. the poor condition of the subsoil or less space for the embankments or the lack of building material and the demand for cost effectiveness, forces to reflect new approaches.

6.2 Practice Integrated Coastal Zone Management

The demand to consider different types of land use in coastal zones was introduced by the concept of Integrated Coastal Zone Management (ICZM). The challenge was to identify and to define what integrated management of a complex system means. The principle of sustainable development is based on three pillars: society, the economy and ecology. Additionally, with the Agenda 21 processes the participation approach was introduced and took on a prominent role in the achievement of sustainable development. Six aims are important for participation processes (Grunwald 2002):

- Enhancement of the knowledge base for decision-making (extension of scientific knowledge by local and experienced knowledge).
- Enhancement of the value basis to widen the social dimensions of decisions.
- Information function, to enable informed evaluation on the part of the public.
- Increase social compatibility through the consideration of different aspects.
- Conflict prevention and resolution by collaboration on joint solutions.
- Orientation on joint welfare, to overcome particular interests through rational discourse.

In line with these aims, the existing instruments can be tested as to whether they are ready for the new challenges or not. With regard to the currently revised spatial planning instruments in Lower Saxony it can be stated that the basis for introducing participation is in place. The cooperation between different sectors, organisations and target groups is envisaged in procedural blueprints. But, there is a gap between theory and practice. Currently, participation is carried out in a formal way, mainly within the spatial planning approval procedures or in environmental impact assessments. New instruments were introduced on the European level which emphasised the involvement of target groups and stakeholders. With regard to coastal protection participation is only possible on a formal basis within the mentioned approval procedures, wherever they become necessary. For example, the Master Plan for Coastal Protection in Lower Saxony (NLWKN 2007b) describes the interlinkages to different items like ICZM. But, for the development of the Master Plan only an institutional consultation process was carried out. And the State Law on Dikes (NDG, MU 2004) does not contain any stipulation on the involvement of different groups or the broad public. The dike boards do not have the tools to conduct participation processes. This applies to both internal and external participation. For the internal part, there might be no demand for an enhanced participation instrument, because the members of the dike boards trust their managing board to take the right decisions for the protection of people and property. The formal procedure of participation is mainly effected through the election of members of a committee which elect the managing board of the dike board. These elections are held every 5 or 6 years. The managing board presents past and future activities in the dike board assemblies. In principle, the option to become actively involved in the dike board as a member is available, in practice this option is seldom taken. And vice versa, the managing board tries to mobilise its members only in special situations, e.g. in the case of the temporary interruption of the coastal protection works at the western part of the Jade Bay (Cäciliengroden), where approx. 10,000 people built a chain of torches on the main dike to demonstrate against this interruption.

For external participation no statutory tool exists, although, the Norden Dike Board ran a discussion process together with the former NLWK about the fore land management in their area of concern – see NLWK Norden (2003). But, the experiences of former coastal protection projects like these in the western part of the Jade Bay and in the Ley Bay lead to cooperative solutions – see e.g. NLÖ (2000). Therefore, the willingness to develop joint solutions may grow within the dike boards. For example, the III. Oldenburg Dike Board has to strengthen the dike in the north part of their area – Elisabeth-Außengroden. The required clay has to be gained many kilometers inland. Consequently, the transportation of the clay will cause a disturbance in a highly frequented tourist and recreational area. The time period for building and strengthening dikes is limited (summer time). So, cooperation is necessary and information on the necessity of the work must be given as well. Furthermore, the dike boards are responsible for removing the flotsam from the dikes, because this can cause serious damages of the grass layer. The dike boards are striving for cooperation with farmers and perhaps biomass power plants to use this material. These approaches are quite new and have to be investigated in more detail, due to the composition of the flotsam, e.g. litter and of course saline plants which can not easily be utilised in biomass power plants. However, this will feature as one aspect of a proposal to the Interreg IVB programme on the utilisation of biomass energy in coastal regions (enerCOAST).

Finally, to some extent approaches do exist and the options are prepared for, but have not really been taken. Spatial planning provides a basis for the transformation of existing coastal protection concepts to spatial protection concepts and thereby comply with the principles of ICZM (see Sect. 4.1.2). The existing strategy and instruments in coastal protection have to be adapted to the new challenges.

6.3 Outcome – Options of Multifunctional Coastal Protection Zones

The long-term perspective on the implementation of spatial coastal protection concepts is one of achieving a sustainable coastal development integrating the requirements of safety against flooding and the desired development of coastal zones. The installation of coastal protection zones as a first step can provide a sound basis for the application of various spatial concepts. The benefits of Multifunctional Coastal Protection Zones (MCPZ) are:

Integration: The integration of coastal protection into spatial planning leads to the insight that coastal protection is a user perspective with distinct spatial interests. The option of developing and testing new coastal protection strategies is already incorporated in spatial plans.

Cooperation: The implementation of spatial coastal protection concepts has the potential to reduce latent land use conflicts in the coastal zone, and thus will lead to an efficient and effective implementation of necessary coastal protection projects.

Multifunctionality: Although a certain area is occupied by coastal protection (as priority or as precautionary area), other types of land use are not completely prohibited, but have to be adapted to the existing circumstances, e.g. excavation of building material. Other structures designed for other functions can take on the function of a second dike line.

Safety: The existing single line of defence can be enhanced by the recommended additional safety element in the hinterland – see e.g. Petersen (1966) and Führböter (1987).

These are the benefits of MCPZ, but how can these zones be implemented? The existing instruments in spatial planning offer the opportunity, but the practical implementation procedure has to be found. The Participatory Integrated Assessment (PIA) process in Nessmersiel provides strong evidence of a viable approach. The benefits of the applied PIA process to initiate a MCPZ are:

Participation: Relevant stakeholders are involved in each step of the entire process.

Vision: The application of the scenario-technique offers the opportunity to anticipate future development and to adequately create adaptation strategies.

Transparency: The involvement of relevant stakeholders throughout the process leads to a maximum of transparency.

Traceability: The introduction of the scoring matrix serves as a discussion platform which provides a basis for the deliberation and negotiation of Design Elements (DE) and the different land use options.

Decision Support: The application of the outranking method offers opportunities to conduct a comprehensive multi-criteria analysis which takes different value categories into account.

Adaptability: The entire process is build on flexible adaptation to different challenges: new demands in the future and revision under changing circumstances.

Sustainability: The demand for cross-sectoral thinking and the integration of a variety of expert-knowledge comply with the requirements of sustainable coastal development.

The different fields of action and the appropriate time-scales for different tasks connected to the implementation of spatial coastal protection concepts have been exemplarily treated in Jeschke (2004), adapted in Klenke et al. (2006) and are shown by Fig. 6.1. This diagramme has to be transferred to a stepwise approach. Flügel and Dziatzko (2008) prepared a sound basis for the options of incorporating coastal protection zones into the spatial planning system of Lower Saxony. The first step concerning spatial planning has been done by the incorporation of spatial coastal

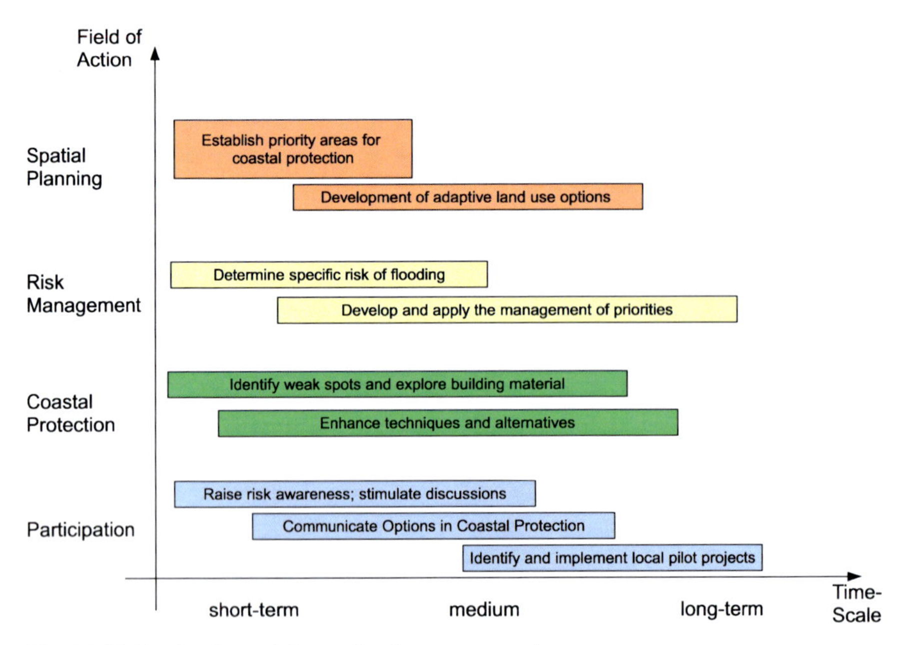

Fig. 6.1 Fields of action and time-scales for necessary tasks
Source: Jeschke (2004) and Klenke et al. (2006).

protection concepts in the LROP of Lower Saxony (ML 2008). The current weak spots have already been described in the latest Master Plan for Coastal Protection (NLWKN 2007b). The next important step will be to determine the priority areas for the coming years (following the current project in the eastern part of the Jade Bay). To start early with the identification and creation of sound sustainable coastal development options (communication and cooperation): this dissertation provides a guide for both the development and the implementation of spatial coastal protection concepts.

6.4 Final Remarks

This dissertation does not follow the usual structure of scientific research. Each chapter is completed with concluding remarks. Thus, it may not be necessary to repeat the conclusions and to recapitulate the recommendations. The fundamental points are communication and cooperation on and between different levels, sectors and branches. The fear of loosing control is widely spread, and has been published as one of the barriers to participation (e.g. Renn et al. 2005): the fear articulated is that power and decisions will be transpired out of the existing spheres of influence, if participation were to be implemented. But, the experiences of the ComCoast project in the UK Abbotts Hall (farm diversification) or Wallasea (wetland creation and farm diversification), in The Netherlands Perkpolder (regional development and compensation) or Ellewoutsdijk (preserve cultural heritage and overtopping resistant embankment) provided evidence that a slow change in thinking is underway (detailed description in Appendix B). The quote of a participant of the Nessmersiel process summarised the situation as follows: "First, I was sceptical if such a discussion process can lead to well-founded planning, but afterwards I can conclude that the participation was no waste of time and I have gained insight into boundary conditions of different sectors".

The Nessmersiel process is encouraging in that it has demonstrated that methodical rigour and mobilisation of diverse resources can generate real perspectives for a sustainable future.

Chapter 7
Summaries

Contents

7.1 Summary

The purpose of this dissertation was twofold: first, to elaborate the options provided by spatial coastal protection concepts for safety and sustainable coastal development, i.e. Multifunctional Coastal Protection Zones (MCPZ) and second to develop a procedure for the implementation of MCPZ, i.e. a Participatory Integrated Assessment (PIA) process.

People settled in the low-lying and flood-prone areas of the southern North Sea. Protection against tidal waters and storm surges was based historically on intuition and experiences. The development of technical constructions to protect inhabitants against the flood started with dwelling mounds and resulted in a continuous single line of defence, the main dike line. This concept is still valid in Lower Saxony and The Netherlands. In addition, further embankments and technical constructions such as sluices, locks and barriers have been installed. After the experiences of severe storm surges in the middle-ages and especially in the last century, the strategy and concepts of coastal protection have been improved and partially revised. In Lower Saxony the design water level was introduced as a technical parameter to calculate the necessary height of dikes. The application of this technical parameter ensures the same safety level along the entire coast line of Lower Saxony. The approach of The Netherlands was to integrate different aspects (spatial quality) in the determination of the safety standard. This has led to different safety levels along the Dutch coast.

A consequence of the percolating environmental issues and the climate change discussions was the installation of the Intergovernmental Panel on Climate Change (IPCC) by the UN. Since then, two important issues appeared on the international and political agenda: climate change and sustainable development. Sustainable development demands an integrated management approach, especially in coastal

F. Ahlhorn, *Long-Term Perspective in Coastal Zone Development*,
DOI 10.1007/978-3-642-01774-2_7, © Springer-Verlag Berlin Heidelberg 2009

zones, i.e. Integrated Coastal Zone Management (ICZM). First, climate change was seen as a natural phenomenon. But today, the thorough and comprehensive results of scientific research indicate that climate change is undoubtedly taking place. The results provide evidence of a (strong) human influence. The likely impacts pose threats to all branches and fields of life, especially at the coasts. The forecasting of the intensity of storms, sea level rising or precipitation is crucial for the calculation of parameters for coastal protection. These aspects contain uncertainties, and thus can only be estimated. However, the principle of sustainable development demands taking future development into account. The integration of different interests and needs and the necessary planning of future development requires adequate methods and tools.

Participation and evaluation have prominent roles within sustainable development processes. Today, both have been featured in many laws, ordinances and plans, but the application of these methods in coastal protection projects is seldom. The question remains as to how to achieve the goals of sustainable coastal development, i.e. the integration of different types of land use in conjunction with the required level of protection against flooding under changing increasingly critical circumstances? Strong evidence was collected to show that the proposed Participatory Integrated Assessment (PIA) process meets this requirement. Participants in the featured process included representatives of e.g. regional and local government, dike board, tourism, nature conservation and coastal protection. The process started with the provision of basic settings, i.e. the description of a plausible future development for the investigation area and the relevant sectors. The adaptation of these settings to local circumstances was made by the participants. On the basis of these settings the participants jointly created integrated scenarios. These integrated scenarios were implemented in a Geographic Information System (GIS) using Design Elements (DE's). Design Elements are spatial units of intermediate scale, positioned between regional and town/urban planning. Each DE has attributes like costs, benefits and impacts, and belongs to one of two value categories: benefit value or functional value. Together with the existence value, these value categories are able to adequately depict sustainable development processes, i.e. integration of societal, economic and ecologic aspects. For example, the salt marsh is a natural unit which provides specific functions for different types of land use, e.g. nature conservation, coastal protection and agriculture. The functional values posed by a salt marsh e.g. for nature conservation is one of information and for coastal protection it serves the regulation of wave energy.

The aim of the evaluation method (outranking) was the comparison of three integrated scenarios in the light of sustainable development. To obtain the relevance of each DE a scoring matrix was introduced. The result of the scoring matrix was the consensus on the preferred land use option in the future. All participants discussed the relevance of a specific DE according to a certain type of land use, e.g. nature conservation or agriculture. A special feature of the scoring matrix was that DE's which were relevant for more than one type of land use were weighted higher and thus were more relevant than others. This feature calls for the multifunctional use of DE's. The result of the outranking method was that the integrated scenario "C" won against the other two which strongly emphasised communication and cooperation and displayed a more sustainable tendency. The participants expected "A" to win, but accepted the result as a consensus of the entire group. Every participant on the potential of this process to support the resolution of land use conflicts and to create visionary land use patterns. Furthermore, the process can be adapted to new tasks in the future. For example, if the insufficient safety level of an embankment requires building material from the neighbourhood.

The Nessmersiel process is encouraging in that it has demonstrated that methodical rigour and mobilisation of diverse resources can generate real perspectives for a sustainable future.

7.2 Zusammenfassung

Das Ziel dieser Dissertation ist die Entwicklung und Erprobung eines Bewertungs- und Beteiligungsprozesses zur Integration von flächenhaften Küstenschutzkonzepten in die räumliche Planung. Insbesondere geht es um die nachhaltige Entwicklung tief liegender Küstengebiete durch die Schaffung "Multifunktionaler Küstenschutzzonen". Mit der Einführung des Begriffes "Nachhaltigkeit" oder "nachhaltige Entwicklung" (sustainable development) wurden neue Anforderungen an die bestehenden Methoden für z.B. die räumliche Planung formuliert. Dem Begriff "Nachhaltigkeit" unterliegt die Idee, dass heutige Entscheidungen die Entwicklungsmöglichkeiten zukünftiger Generationen nicht übermäßig beeinträchtigen sollen. Dieser Idee nachzukommen, würde bedeuten die zukünftige Entwicklung in heutigen Entscheidungen zu berücksichtigen. Die Ansätze zukünftige Entwicklungen in heutige Entscheidungen einzubeziehen, finden auf unterschiedlichen Ebenen statt, z.B. der Beteiligung von Personen oder Betroffenen (Partizipation) oder der Bewertung von sozio-ökonomischen, ökologischen und gesellschaftlichen Aspekten. Mit der Einführung des Begriffes der Nachhaltigkeit wurde grundsätzlich eine breite Einbindung aller Betroffenen an Entscheidungsprozessen gefordert (Agenda 21). Darüber hinaus wird Nachhaltigkeit als die ausgewogene Berücksichtigung der drei Pfeiler Gesellschaft, Ökonomie und Ökologie verstanden (Integration). In nachhaltigen Entscheidungsprozessen sollten diese drei Pfeiler gleichrangig in einen Bewertungsprozess eingebunden werden, auf dessen Grundlage eine Entscheidung getroffen werden kann. Auf der Basis dieser grundlegenden Ideen wurde der Bewertungs- und Beteiligungsprozess für die Schaffung "Multifunktionaler Küstenschutzzonen" entwickelt und erprobt: Einbindung aller relevanten Betroffenen (Beteiligung) und Schaffung einer Grundlage für einen nachhaltigen Entscheidungsprozess (Bewertung).

Ausgangspunkte für die Bearbeitung dieser Thematik sind die Erkenntnisse und Aussagen des vierten Berichts des Intergovernmental Panel on Climate Change (IPCC) zu den Auswirkungen des Klimawandels auf Küstenregionen, den damit verbundenen Anforderungen an den Küstenschutz und eine räumliche Planung für die Zukunft. In dieser Arbeit steht nicht die Frage nach dem Ausmaß des menschlichen Einflusses auf den Klimawandel im Vordergrund, sondern die möglichen Konsequenzen eines Klimawandels auf die Strategien des Küstenschutzes im südlichen Nordseeraum. Für tiefliegende Küstenregionen, wie dem südlichen Nordseeraum, spielen die Höhe und Geschwindigkeit eines Meeresspiegelanstieges (Sea Level Rise) sowie die Veränderung der Sturmtätigkeit eine große Rolle. Bis zu den verheerenden Sturmfluten 1953 in den Niederlanden und 1962 in Norddeutschland basierte der Deichbau auf den tradierten Erfahrungen der Menschen aus den vergangenen Jahrhunderten. Über Generationen hat der Küstenschutz im südlichen Nordseeraum auf höher auflaufende Sturmfluten mit der Erhöhung, Verstärkung und Verkürzung der Hauptdeichlinie reagiert. Konsequenz dieser Strategie ist der steigende Bedarf an Raum und Material für den Deichbau. Darüber hinaus sind zunehmende Raumnutzungsansprüche auf Grund der ökonomischen Entwicklung in den besiedelten Küstenregionen zu berücksichtigen.

In der Raumplanung wird das Instrumentarium für die notwendigen planerischen Vorgänge in den verschiedenen Planungsebenen (in Deutschland: Bund, Land, Region, Kommune) bereitgestellt, um eine planmäßige Ordnung, Entwicklung und Sicherung zur bestmöglichen Nutzung des Lebensraumes zu gewährleisten. Die Raumplanung soll vermehrt dem Prinzip der Nachhaltigkeit genügen, das seit 1992 in Rio de Janeiro eine integrierte Betrachtung speziell für die Küstenregionen fordert, das so genannte Integrierte Küstenzonenmanagement (IKZM). Trotz intensiver Diskussionen und Forschungen ist noch kein einheitliches Konzept zum IKZM entstanden. Auf europäische Ebene wurden mit den Empfehlungen zu einem IKZM grundlegende Prinzipien eingeführt (EC 2002). Diese Prinzipien erweitern die Inhalte und Intentionen der bisherigen raumplanerischen Instrumente, in dem sie die Nutzung von Szenarien, die Beteiligung aller Betroffenen (informell, nicht nur formal) oder eine integrierte, also sektorenübergreifende, Planung fordern.

Die Ergebnisse des Bewertungs- und Beteiligungsprozesses zur Umsetzung "Multifunktionaler Küstenschutzzonen" wird anhand von drei Fragen erläutert:

1. Was ist eine "Multifunktionale Küstenschutzzone"? Die Grundlage für die Umsetzung einer "Multifunktionalen Küstenschutzzone" bildet die in dieser Frage gegebene Definition.
2. Sind die raumordnerischen Voraussetzungen für "Multifunktionale Küstenschutzzonen" vorhanden? Die Möglichkeit der Umsetzung einer "Multifunktionalen Küstenschutzzone" im niedersächsischen Küstenraum wird anhand dieser Frage erläutert. und
3. Wie ist die Umsetzung "Multifunktionaler Küstenschutzzonen" möglich? Die Antwort auf diese Frage ist das Ergebnis des Bewertungs- und Beteiligungsprozesses aus der Fallstudie Neßmersiel.

Die Fallstudie "Neßmersiel" wurde im Rahmen des europäischen INTERREG IIIB Projektes ComCoast (*Com*bined Functions in *Coast*al Defence Zones) vom deutschen Partner dem Institut für Chemie und Biologie des Meeres (ICBM), Arbeitskreis IKZM, bearbeitet.

Zur Hinführung auf die Definition einer "Multifunktionalen Küstenschutzzone" wird kurz auf die Entwicklung des Küstenschutzes im südlichen Nordseeraum eingegangen. Im südlichen Nordseeraum hat der linienhafte Küstenschutz eine über 1000-jährige Tradition. Die durch die Hauptdeichlinie in Niedersachsen geschützte Fläche beträgt etwa 6,900 km^2, auf der ungefähr 1,2 Mio. Menschen leben und arbeiten. Für die Niederlande umfasst die geschützte Fläche ungefähr zwei Drittel des gesamten Landes mit etwa 7 Mio. Einwohnern. Mit der fortwährenden Verbesserung des Schutzes der tiefliegenden Küstenregionen durch Deiche und Dämme ist die Nutzung des Raumes immer weiter intensiviert worden. Ausbau von Siedlungen und industriell genutzter Fläche, aber auch die landwirtschaftliche und touristische Nutzung haben immer mehr zugenommen. In den 70ern des letzten Jahrhunderts hat der Natur- und Umweltschutz zunehmend an Bedeutung gewonnen. In Niedersachsen durch die Einrichtung des Nationalparks "Niedersächsisches Wattenmeer" im Jahr 1986 und vielen Natur- und Landschaftsschutzgebieten hinter den Deichen. Die Raumordnung schafft den Rahmen und die Vorgaben für die Entwicklung und Sicherung von Regionen. In tief liegenden Küstenregionen sichert der Küstenschutz den Lebens- und Wirtschaftsraum. Der Küstenschutz im südlichen Nordseeraum prägt durch sein linienhaftes Schutzkonzept die raumplanerischen Vorgaben. Diese raumordnerischen Pläne und Programme enthielten bisher keine strategischen oder konzeptionellen (raumbezogenen) Aussagen zum Thema Küstenschutz. Die Umsetzung flächenhafter Schutzkonzepte bedürfen der Berücksichtigung in diesen raumordnerischen Plänen und Programmen.

Die größte räumliche Einheit bildet das "Küstenschutzgebiet", darin enthalten ist die "Küstenschutzzone" und diese wiederum beinhaltet den "Küstenschutzstreifen". Das "Küstenschutzgebiet" umfasst die gesamte überflutungsgefährdete Fläche. Diese Fläche wird landseitig vom Geestrand mit +8 m NN Geländehöhe (Anhang zum Niedersächsischen Deichgesetz, NDG) begrenzt und reicht seeseitig bis zur −10 m Tiefenlinie, in der noch morphologisch wirksame Prozesse für den Küsten- und Inselschutz stattfinden. Die "Küstenschutzzone" wird seeseitig durch den im NDG definierten Sicherheitsstreifen (§23, erwünschte Vorlandbreite von 200 m; wenn nicht vorhanden, dann ist ein 500 m Wattstreifen für den Küstenschutz zu bestimmen) begrenzt. Landseitig endet die "Küstenschutzzone" an einem Element (§29 NDG), welches geeignet ist, bei Versagen der Hauptdeichlinie das Hinterland vor Überflutungen zu schützen oder dieses zumindest zu verzögern (z.B. 2. Deichlinie). Die kleinste räumliche Einheit bildet der "Küstenschutzstreifen", dessen seeseitiger Anteil durch den zuvor genannten Sicherheitsstreifen gegeben ist. Landseitig ist der "Küstenschutzstreifen" durch einen, ebenfalls im NDG genannten, 50 m breiten Puffer abgegrenzt (§16 NDG). Die Multifunktionalität ist über die gleichzeitige Nutzung dieser räumlichen Einheiten durch andere Raumnutzer wie Landwirtschaft, Naturschutz, Wasserwirtschaft gegeben. Durch die Definition räumlicher Einheiten für den Küstenschutz, ist dieser explizit als "Raumnutzer" ausgewiesen und mit entsprechenden Ansprüchen in die raumplanerischen Pläne und Programme zu integrieren.

Damit stellt sich die Frage, inwieweit die raumordnerischen Voraussetzungen für "Multifunktionale Küstenschutzzonen" vorhanden sind. Mit der Neufassung des Landesraumordnungsprogramms (LROP) des Landes Niedersachsen von 2008 werden entsprechende Aussagen zum Küstenschutz aufgenommen: Zum einen soll "vor dem Hintergrund zu erwartender Klimaveränderungen die Erforschung, Entwicklung und Erprobung alternativer Küstenschutzstrategien Rechnung getragen werden" (LROP: Kap. 1.4 Abs. 12). Zum anderen sind in den nachgeordneten Regionalen Raumordnungsprogrammen (RROP) vorsorgend Flächen für Deichbau und Küstenschutzmaßnahmen zu sichern (LROP: Kap 3.2.4 Abs. 10). Bei Maßnahmen des Küsten- und Hochwasserschutzes sollen die Belange der Siedlungsentwicklung, der Wirtschaft, des Naturschutzes, der Landschaftspflege, des Tourismus und der Erholung berücksichtigt werden. Zusätzlich wurde das Planungsgebiet des LROP auf die 12 Seemeilenzone ausgedehnt, womit erstmals die Land-Meer-Grenze überschritten wurde und eine integrierte Planung für die Küstengewässer und die Landfläche möglich wird. Diese Neuerungen müssen allerdings noch in die zum Teil in der Überarbeitung befindlichen RROP übernommen werden. Die Antwort auf die oben gestellte Frage lautet: Ja, die raumordnerischen Voraussetzungen sind auf der Ebene der Landesplanung gegeben, diese müssen aber noch in die RROP eingearbeitet werden. Aus der Sicht des Küstenschutzes sind Bausteine für flächenhafte Schutzkonzepte zum Teil gegeben (siehe Definitionen oben), doch bedarf es einer konzeptionellen Änderung des Ansatzes zur Bestimmung des erforderlichen Sicherheitsstandards von Schutzelementen, vom deterministischen zum probabilistischen Ansatz. Die Bestimmung des Bemessungswasserstandes und des Wellenauflaufs, mit deren Hilfe die Höhe eines Deiches ermittelt wird, setzt sich bis heute aus der Addition verschiedener Einzelparameter zusammen (Einzelwertverfahren). Ein probabilistischer Ansatz würde die Wahrscheinlichkeit des Eintretens einzelner Parameter (Höchster Wasserstand, höchster Wellenauflauf) für die Ermittlung der Deichhöhe berücksichtigen. Dieser Ansatz müsste in ein umfassenderes Risikomanagement für tief liegende Küstengebiete eingebettet werden, in dem neben der Versagenswahrscheinlichkeit eines Bauwerkes auch die dahinter liegenden Sachwerte und das zu akzeptierende Restrisiko betrachtet werden müssen. Eine ausführliche Beschreibung ist in den vorhergehenden Kapiteln zu finden.

Für die Beantwortung der dritten Frage, wie die Umsetzung einer "Multifunktionalen Küstenschutzzone" möglich ist, werden zunächst die notwendigen Anforderungen an einen integrierten Bewertungs- und Beteiligungsprozess erläutert. Diese Anforderungskataloge sind in internationalen, europäischen und nationalen Plänen, Programmen und Übereinkommen enthalten, beispielweise in der Erklärung von 1992 über das Prinzip der Nachhaltigkeit, speziell zum IKZM. Auf europäischer Ebene wären z.B. die Aarhus Konvention für den freien Zugang zu Informationen oder die EU Empfehlungen von 2002 zu einem IKZM zu nennen. Zusammengenommen geht es hauptsächlich um die Partizipation, was eine umfassende Bedarfsanalyse der relevanten Beteiligten und Betroffenen (Stakeholder) einbezieht sowie deren umfangreiche Beteiligung im jeweiligen Entscheidungsprozess. Darüber hinaus wird empfohlen, objektive Bewertungsmethoden einzuführen, welche sowohl sozio-ökonomische als auch ökologische Aspekte berücksichtigen, um dem Prinzip der nachhaltigen Entwicklung zu entsprechen. Diese Prozesse sollen transparent und nachvollziehbar sein und lang- wie kurzfristige Planungshorizonte abdecken. Die eingeführten Bewertungsmethoden sollen der Entscheidungsunterstützung dienen und, da es sich um räumliche Planungsprozesse handelt, die Möglichkeit bieten, räumliche Kriterien zu berücksichtigen.

Der Schwerpunkt dieser Dissertation liegt in der Entwicklung und Erprobung dieses Bewertungs- und Beteiligungsprozesses unter der Berücksichtigung der zuvor genannten Anforderungen. Das Ziel für die Fallstudie "Neßmersiel" war es, eine multifunktionale Raumnutzung unter der Maßgabe einer nachhaltigen Entwicklung im Küstenraum für das Jahr 2050 zu erarbeiten. Der eingesetzte Methodenrahmen, ein Participatory Integrated Assessment (PIA) Prozess, ermöglicht es, diesen Anforderungen gerecht zu werden. Dieser Methodenrahmen ist bereits in verschiedenen Forschungsbereichen angewendet und erprobt worden. Erstmalig wurde dieser Methodenrahmen in der vorliegenden Dissertation für die Integration des Küstenschutzes in die Raumplanung eingesetzt. Der entwickelte PIA Prozess besteht aus drei wesentlichen Arbeitsschritten. Im ersten Schritt steht die Anwendung der Szenario-Technik und die Einführung räumlicher Kriterien (Design Elemente) im Mittelpunkt. Im zweiten Schritt gilt es, die Gewichtung dieser räumlichen Kriterien zu ermitteln. Und im dritten Schritt wird die Evaluierung des Prozessergebnisses sowie des gesamten PIA Prozesses durchgeführt.

Die Fallstudie "Neßmersiel" wurde im Verbandsgebiet der Deichacht Norden (Ostfriesland), einer Region im nordwestlichen Niedersachsen, durchgeführt. Räumlich begrenzt ist das Untersuchungsgebiet (UG) seeseitig durch das Wattenmeer, beginnend mit der Mittleren Tidehochwasserlinie (MThw-Linie), und landseitig durch die zweite Deichlinie. Das UG besteht aus einem bis zu 600 m breiten Deichvorland, zu dem Salzwiesen und ein Sommerpolder gehören. Der Sommerpolder ist eine ehemalige Salzwiese, die durch einen Sommerdeich gegenüber Windfluten geschützt ist. Daran schließt sich der Hauptdeich an, der das dahinter liegende Gebiet gegen Überflutung schützt. Zwischen dem Hauptdeich und der zweiten Deichlinie befindet sich eine landwirtschaftlich intensiv genutzte Polderfläche. Das gesamte UG umfasst ungefähr 700 ha. Der Sommerpolder wird extensiv durch Beweidung und in den Sommermonaten hauptsächlich von Erholungssuchenden genutzt. Das Deichvorland untersteht dem Schutz des Nationalparks "Niedersächsisches Wattenmeer" mit unterschiedlichen Schutzkategorien (Ruhezone und Zwischenzone). Das an das Deichvorland im Osten angrenzende Hafengebiet dient in der Hauptsache als Fährhafen zur Ostfriesischen Insel Baltrum, der angrenzende Strandbereich liegt in der Erholungszone des Nationalparks.

Der erste Vorbereitungsschritt für das Design des PIA Prozesses war eine umfangreiche Analyse der zu beteiligenden Institutionen und Organisationen. Als Prozessteilnehmer wurden die folgenden Nutzergruppen identifiziert: Küstenschutz (Verband und Behörde), Naturschutz (Verband und Behörde), Landwirtschaft (Kammer), Gemeinde (Planung und Tourismus), Regional- und Landesplanung. Im so genannten "Kick-off Meeting", dem ersten Treffen der Lokalen Kontaktgruppe (LKG), wurden sowohl die Zielsetzung des EU INTERREG IIIB Projektes ComCoast als auch die speziellen Ziele der deutschen Fallstudie erläutert. Mit der Beschreibung und Dokumentation des Status quo des UG erhalten alle

Mitglieder der LKG einen annähernd gleichen Kenntnisstand über das UG. Im zweiten Teil dieses Treffens wurden Grundannahmen für die drei integrierten Szenarien des Bezugszeitraumes (2050) vorgestellt und durch die Mitglieder der LKG an die lokalen Gegebenheiten angepasst. Die Grundannahme für den Küstenschutz war, dass er in allen Szenarien einen hohen gesellschaftlichen Stellenwert einnehmen wird. Die Annahmen für die Faktoren des Klimawandels bis 2050 sind wie folgt: 30 cm Meeresspiegelanstieg, Watten und Vorländer wachsen mit, der Niederschlag wird sich um 10% im Jahresverlauf erhöhen und der Wind wird sich geringfügig ändern. Die morphologische Entwicklung der Vorländer wurde auf der Grundlage beobachteter Trends fortgeschrieben, so dass erosive Tendenzen in unterschiedlicher Ausprägung in den geschützten und ungeschützten Bereichen des Deichvorlandes stattfinden. Darüber hinaus gab es plausible Annahmen zu den Entwicklungen in den Bereichen der ökonomischen und der demografischen Entwicklung, der regionalen Identität sowie Veränderungen in den sozialen und politischen Rahmenbedingungen. Somit basieren diese "integrierten" Szenarien auf plausiblen Annahmen über die Veränderungen gesellschaftlicher, ökonomischer und demografischer Aspekte.

Der erste Arbeitsschritt im PIA Prozess bestand aus der gemeinsamen Erarbeitung von Landnutzungsmustern für das UG im Jahr 2050 unter der Berücksichtigung der Grundannahmen und der drei integrierten Szenarien. Für die erfolgreiche Durchführung des ersten Arbeitsschrittes, der Anwendung der Szenario-Technik, war die Vorbereitung der LKG-Mitglieder notwendig. Diese bestand aus der gedanklichen Auseinandersetzung mit möglichen Reaktionen einer jeweiligen Nutzergruppe auf die Rahmenbedingungen der drei vorgestellten integrierten Szenarien. Für die erfolgreiche Bearbeitung dieses Arbeitsschrittes wurden verschiedene Werkzeuge und Methoden bereitgestellt bzw. entwickelt. Zum Einsatz kam ein Geografisches Informationssystem (GIS), das hervorragend für die räumliche Darstellung von Informationen geeignet ist und über den Einsatz räumlicher Kriterien zusätzlich für das entwickelte Bewertungssystem genutzt wurde. Diese räumlichen Kriterien werden als Design Elemente (DE) bezeichnet und sind in der Raumordnung zwischen den Ebenen der Bauleitplanung und der Regionalplanung angesiedelt. In diesen Planungsebenen werden so genannte Planzeichen eingesetzt, um die räumliche Nutzung darzustellen (blaue Fläche = Wasserfläche). Die DE sind um eine wichtige Eigenschaft gegenüber Planzeichen erweitert worden: Bisherige Planzeichen sind höchstens einer Nutzwertanalyse zugängig, in dem für landwirtschaftliche Flächen über den Erlös pro Hektar Ackerland bewertet werden kann. Dies ist jedoch nicht ausreichend, soll dem Anspruch einer nachhaltigen Entwicklung entsprochen werden. Aus diesem Grund wurden die entwickelten DE den Wertkategorien des Bewertungsansatzes des "Ökonomischen Gesamtwertes" zugeordnet. Innerhalb des "Ökonomischen Gesamtwertes" wird in nutzungsabhängige und nicht-nutzungsabhängige Wertkategorien unterschieden. Diese Wertkategorien werden noch weiter unterteilt, worauf aber in dieser Dissertation nicht weiter eingegangen wurde, da sie den Untersuchungsrahmen übersteigen würden.

Die für die Fallstudie “Neßmersiel” angewendeten Wertkategorien umfassten die Funktions- und Nutzwerte (nutzungsabhängig: indirekt und direkt) sowie den Existenzwert (nicht-nutzungsabhängig). Der Funktionswert beschreibt, die ökologischen Leistungen, die die Biosphäre einem Nutzer zur Verfügung stellt und die sich durch ihn nutzen lassen (für die Landwirtschaft wäre dies z.B. die Ertragsfunktion, für den Küstenschutz die Regulationsfunktion der Salzwiesen in Bezug auf Energiedissipation von Wellen). Der wirtschaftliche Nutzwert ist die für Produktions- oder Konsumzwecke direkt genutzte biosphärische Leistung. Für den Existenzwert steht weder die direkte noch die indirekte Nutzung im Mittelpunkt, das Wissen um die bloße Existenz eines Naturgutes wirkt Wert stiftend. Die DE wurden a priori unterschiedlichen Nutzergruppe zugeordnet. Dem Naturschutz wurden die Pionierzone, die Salzwiese, die Marschboden, das Schutzgebiet, das Informationszentrum, der Salzwiesenlehrpfad und die Information und Beobachtung zugeordnet. Der Funktionswert für die Nutzergruppe Naturschutz ist die “Information”. Die DE Pionierzone, Salzwiese, Marschboden, Schutzgebiet wurden diesem Funktionswert zugeordnet. Wohingegen die DE Informationszentrum, Salzwiesenlehrpfad und Information und Beobachtung dem Nutzwert zugeordnet wurden, da sich ein bestimmter direkter Nutzen mit ihnen verbindet. Auf diese Art und Weise wurden die bereitgestellten DE den jeweiligen Nutzergruppen Küstenschutz, Tourismus, Naturschutz, Energie, Landwirtschaft und Verkehr zugeordnet. Mit Hilfe der DE hatten die Mitglieder der LKG gemeinsam Landnutzungsmuster für das Jahr 2050 erstellt. Diese Landnutzungsmuster stellten den Konsens der Gruppe über die zukünftige Nutzung des UG unter Berücksichtigung der drei integrierten Szenarien dar.

Diese drei Landnutzungsmuster dienten als Eingangsgrößen für den folgenden Arbeitsschritt, der Gewichtung der DE im UG. Grundsätzlich lassen sich die drei Muster in ein konservatives, also der Fortsetzung des Status quo, in ein pessimistisches und ein optimistisches unterscheiden. Der Existenzwert für das UG wurde beschrieben durch die Kriterien “Vorhandensein naturraumtypischer Pflanzen- und Tierarten”, “natürliche Lebensraumstrukturen”, “Kulturlandschaft, Schutzgut Mensch, Infrastruktur” und “Heimat, Schönheit”. Aus den Fragen, die zur Ermittlung des Existenzwertes für das UG entworfen wurden, ließ sich die Aussage treffen, dass die Teilnehmer das Eintreten des konservativen Landnutzungsmusters als wahrscheinlich ansahen. Dem optimistischen Landnutzungsmuster wurde aber der höchste Stellenwert bezogen auf das Vorhandensein der oben genannten Kriterien des Existenzwertes zuwiesen.

Die Herausforderung bei der Gewichtung war, die mehrfache Nutzung von DE zu stimulieren und dies über die Berechnung der Gewichte in den gesamten Bewertungsprozess einfließen zu lassen. Die in dieser Dissertation entwickelte Gewichtungsmatrix erfüllt zusammen mit der angewendeten Bewertungsmethode diesen Anspruch. Die einzelnen Nutzergruppen wurden als Spaltenüberschriften in die Gewichtungsmatrix eingetragen. Die Vergabe von Punkten für jede Nutzergruppe (Wichtigste = 100 Punkte) beschrieb grob, welches Landschaftsbild im UG vorherrschen soll. Im Anschluss wurde die Relevanz eines DE bezüglich der entsprechenden Nutzergruppe bewertet: Das wichtigste DE erhielt pro Nutzergruppe

100 Punkte, alle anderen relativ dazu weniger. Entscheidend war, dass DE, die in einer Nutzergruppe Punkte erhalten, auch für andere Nutzergruppen Punkte erhalten konnten und somit an Relevanz gewannen. So konnte beispielweise für die Nutzergruppe Naturschutz das Informationszentrum wichtig sein. Aber auch die Nutzergruppe Landwirtschaft kann ein Informationszentrum als wichtig erachten, weil gemeinsam über landwirtschaftliche und naturschutzfachliche Kooperationen berichtet werden kann. Somit drückten die Punkte, die ein DE von anderen Nutzergruppen erhalten hat, einen gemeinsamen Vorteil für die Einrichtung eben dieses DE aus. Im nächsten Schritt wurden die Präferenzmatrizen der DE mit der Gewichtung verknüpft. Dabei musste die eingesetzte Bewertungsmethode in der Lage sein, sowohl ordinale als auch nominale Werte zu verarbeiten. Die Bewertung, welches der drei Landnutzungsmuster eine nachhaltigere Entwicklung im Küstenraum darstellt, wurde mit Hilfe eines Outranking Verfahrens ermittelt. PROMETHEE (Preference Ranking Organisation Method for Enrichment Evaluation) erfüllt diese Vorgaben und arbeitet wie folgt.

Das Verfahren basiert auf der Annahme, dass multikriterielle Bewertungen nicht zufriedenstellend durch die gleichzeitige Optimierung aller Kriterien einer Nutzwertfunktion gewährleistet werden können. In aufgestellten Nutzwertfunktionen ist eine differenzierte Optimierung eines jeden Kriteriums schwierig. Aus diesem Grund wurden in den 80er Jahren des letzten Jahrhunderts die Outranking Verfahren entwickelt, die es erlauben, jedes Kriterium gesondert zu betrachten.

Die von den Mitgliedern der LKG erarbeiteten drei Landnutzungsmuster bestehen aus den räumlichen Kriterien (Design Elemente) in unterschiedlicher Ausprägung. Das Informationszentrum beispielsweise ist in zwei von drei Landnutzungsmustern vorhanden und erbringt unterschiedliche Erlöse. Die intensiv genutzten Ackerflächen haben in den drei Mustern eine unterschiedliche Größe und damit werden unterschiedliche Erträge auf den Flächen erwirtschaftet. Der Sommerdeich ist in einem der Muster vorhanden, in einem anderen ist er an drei Stellen durchbrochen und im dritten vollständig abgebaut worden. Dazu gibt es verschiedene Strategien der Vorlandnutzung und des Vorlandschutzes durch verschiedene Raumnutzer. Diese unterschiedlichen Herangehensweisen haben Auswirkungen auf die eingesetzten finanziellen Mittel (z.B. Kosten für den Küstenschutz). Auch die Nutzung der Polderflächen unterscheidet sich in den Landnutzungsmustern (Höhe der erwirtschafteten Erträge).

Die unterschiedlichen Ausprägungen einzelner DE (Höhe der Erträge pro Ackerfläche, Erlös aus dem Informationszentrum, Kosten für den Küstenschutz) wurden von der "Multikriteriellen Matrix" des PROMETHEE Verfahrens wiedergespiegelt, in dem für jedes Landnutzungsmuster unterschiedliche Werte (nominal, ordinal, kardinal) pro Kriterium ermittelt wurden. Im folgenden Schritt wurde pro Kriterium ein Vergleich zwischen den drei verschiedenen Landnutzungsmustern erstellt. Dies wurde ebenfalls für den Existenzwert durchgeführt. In der Fallstudie "Neßmersiel" wurde für diese Vergleichsoperation die einfache Präferenz angewendet. Die Entscheidung über die , Maximierung oder Minimierung eines Kriteriums wurde mit den Mitgliedern der LKG diskutiert. Das Ergebnis ist eine "Präferenzmatrix" für ein spezielles Kriterium. Die Präferenzmatrizen werden mit den ermittelten Gewichten multipliziert. Die Verknüpfung aller Präferenzmatrizen ergibt die "Gesamtpräferenzmatrix". In der Gesamtpräferenzmatrix spiegelt sich die Präferenz des jeweiligen Landnutzungsmusters wieder.

Als letzter Schritt im PIA Prozess erfolgte die Diskussion des Prozessergebnisses sowie des gesamten Bewertungs- und Diskussionsprozesses mit den Teilnehmern. Die Vorstellung des Ergebnisses der Gesamtpräferenzmatrix hat die Mitglieder der LKG zunächst überrascht. Die Mehrzahl der Mitglieder hatte erwartet, dass das konservative Landnutzungsmuster präferiert werden würde (siehe Existenzwert). Die Erläuterung, warum in der Reihenfolge das optimistischere Landnutzungsmuster vor dem konservativen und dieses wiederum vor dem pessimistischen Landnutzungsmuster liegt, verdeutlichte die Arbeitsweise des Verfahrens und auch dessen Vorteile. Einerseits steigt für Kriterien mit einer mehrfachen Nutzung durch verschiedene Nutzerperspektiven die Relevanz und damit die Gewichtung. Andererseits sind im optimistischeren Landnutzungsmuster DE vorhanden, die in den anderen Mustern nicht vorhanden sind. Mit dem hier beschriebenen Methodenrahmen wird eine nachhaltige Entwicklung des Raumes durch die Berücksichtigung verschiedenster Aspekte (sozioökonomischer und ökologischer) im PIA Prozess unterstützt. Der Einsatz der Gewichtungsmatrix stimulierte zum einen die multifunktionale Nutzung von DE. Zum anderen wurden die DE den verschiedenen Wertkategorien (Funktions-, Nutz- und Existenzwert) zugeordnet. Neben dem reinen wirtschaftlichen Nutzwert, ist der Funktionswert in das Bewertungssystem eingeflossen. Der Funktionswert schätzt den Wert zur Verfügung stehender biosphärischer Leistungen für verschiedene Raumnutzer ab und berücksichtigt darüber hinaus den schonenden und funktionserhaltenden Umgang mit natürlichen Ressourcen.

Die Mitglieder der LKG zogen ein positives Fazit aus dem Ergebnis des PROMETHEE Verfahrens, da die Grundlagen dieser Bewertung, die Landnutzungsmuster und die Gewichtung, gemeinsam und im Konsens erarbeitet wurden. Aus diesem Grund können sich alle Mitglieder auch im Endergebnis wiederfinden, da es transparent und nachvollziehbar ist. Das Verfahren ist anpassungsfähig, denn eine Überprüfung der Präferenzen bzw. die Berücksichtigung aktueller Entwicklungen in bestimmten zeitlichen Abständen ist leicht möglich. Die Übertragbarkeit des Verfahrens auf andere Regionen bzw. Räume kann durch die Anpassung der Liste der DE vollzogen werden. Die Arbeitsschritte des PIA Prozesses können in gleicher Form wieder durchlaufen werden. Somit kann als abschließendes Fazit dieses Bewertungs- und Diskussionsprozesses festgestellt werden, dass das Ergebnis von den Mitgliedern der LKG als gemeinsame Vision für das Jahr 2050 anerkannt worden ist.

Damit kann auch die letzte Frage, also wie die Umsetzung einer "Multifunktionalen Küstenschutzzone" möglich ist, beantwortet werden: Die Durchführung des oben beschriebenen Participatory Integrated Assessment (PIA) Prozesses ermöglicht die Umsetzung einer "Multifunktionalen Küstenschutzzone". Dabei müssen zwei wichtige Aspekte berücksichtigt werden: Die Bearbeitung muss auf räumlichen Kriterien (so genannten Design Elementen) basieren und die

Multifunktionalität muss durch die Gewichtung stimuliert werden. Für beide Aspekte wurden in dieser Dissertation neue Werkzeuge und Konzepte am Beispiel der Fallstudie "Neßmersiel" erarbeitet, angewendet und diskutiert. Welche Bedeutung hat das hier vorgestellte und diskutierte Verfahren für die integrierte Planung? Eine multifunktionale Nutzung von DE wird zu einer nachhaltigen Entwicklung führen. Innerhalb raumplanerischer Verfahren, beispielsweise für Infrastrukturprojekte, sollte über den sektoralen Bedarf hinaus nach Möglichkeiten der zusätzlichen Nutzung gesucht werden: So können höher liegende Straßen als zweite Deichlinien aber auch als Evakuierungsstraßen im Katastrophenfall dienen. Diese Eigenschaften sollten bei der Planung entsprechender Projekte von Beginn an berücksichtigt werden, um mögliche Synergieeffekte zu erzielen. Hierfür liefert das in der vorliegenden Dissertation entwickelte Verfahren entsprechende Methoden und Werkzeuge.

Am Schluss dieser Dissertation stellt sich eine weitere Frage, die hier beantwortet werden soll: Welcher innovative Ansatz ist mit der Umsetzung "Multifunktionaler Küstenschutzzonen" für den Küstenschutz verbunden? Zu Beginn dieser Zusammenfassung wurden die Herausforderungen an die traditionelle Herangehensweise des Küstenschutzes beschrieben. Die Fokussierung auf eine Linie, den Hauptdeich, hat Vorteile für die Instandhaltung dieser Schutzlinie. Diese Erfahrungen sind aus den letzten Jahrhunderte erwachsen. Der stattfindende Klimawandel beeinflusst in einem noch nicht hinreichend geklärten Umfang die Bemessungsgrundsätze für diese Schutzlinie, wie Extremwasserstände und Seegangsklima. Damit sind in diesen Bemessungsgrundsätzen Unsicherheiten enthalten, die es zu quantifizieren gilt. Das Risikomanagement bestehend aus der Risikoanalyse, der Risikobewertung und der Umsetzung von Maßnahmen zur Reduktion des Risikos, versucht diese Unsicherheiten in einen breiteren Kontext einzubetten. Dazu gehören beispielsweise die Abschätzung eines Extremereignisses (Sturmflut) und den damit verbundenen Konsequenzen (Versagen von Bauwerken, Schäden). Aus technischer Sicht wird das Risiko durch die Versagenswahrscheinlichkeit eines technischen Bauwerkes und das Schadenspotenzial des dahinter liegenden Raumes ermittelt. Das würde für den heutigen Bemessungsansatz bedeuten, dass bei gleicher Sicherheit (und damit gleiche Versagenswahrscheinlichkeit des technischen Bauwerkes) ein unterschiedliches Risiko an der niedersächsischen Küste vorhanden ist: Ländlich geprägte Gebiete unterscheiden sich von urbanen Gebieten durch die vorhandenen (Sach-)Werte. Darüber hinaus führt die fortschreitende Entwicklung zu einem Anstieg dieser (Sach-)Werte. Um die Versagenswahrscheinlichkeit eines Bauwerkes zu bestimmen, müssen neben den genannten Bemessungsgrundsätzen auch lokale Gegebenheiten wie die Tragfähigkeit des Untergrundes einbezogen werden. Hinzu kommen Forderungen, die mit dem Prinzip der nachhaltigen Entwicklung verbunden sind, die Berücksichtigung gesellschaftlicher, ökonomischer und ökologischer Aspekte.

Das hier vorgestellte Konzept der "Multifunktionalen Küstenschutzzonen" bietet dem Küstenschutz die Möglichkeit, sich auf ändernde Ansprüche anderer Raumnutzer oder an neue, gesicherte Erkenntnisse der Auswirkungen des Klimawandels flexibel anzupassen. Ein flächenhaftes Küstenschutzkonzept bedarf aber auch der Anpassung durch die jeweiligen Raumnutzer innerhalb dieser Küstenschutzzonen. In der Fallstudie "Neßmersiel" beispielsweise wurde innerhalb des optimistischen Landnutzungsmusters der Hauptdeich nicht traditionell erhöht und verstärkt. Die Grasnarbe auf der Binnenseite des Hauptdeiches wurde mit einer geotextilen Schicht verstärkt und kann so einen größeren Wellenüberlauf und damit einem Überströmen länger standhalten. Als Folge muss das Wassermanagement in der dahinter liegenden Fläche sowie z.B. die landwirtschaftliche Nutzung angepasst werden. Im Rahmen der Fallstudie "Neßmersiel" wurden Versuchsflächen für salztolerante Pflanzen im Polder eingerichtet.

Die Gewinnung des Baumaterials für den Hauptdeich ist an vielen Stellen im südlichen Nordseeraum ein Problem. In einem raumbezogenen Ansatz, wie er in der Fallstudie "Neßmersiel" fiktiv geplant wurde, und wie ihn die Aussagen des LROP in Niedersachsen ermöglichen, können entsprechende Flächen für den Küstenschutz als Vorsorge- oder Vorranggebiet ausgewiesen werden. Aus raumordnerischer Sicht bedarf dies aber einer noch weitergehenden Untersuchung. Dies betrifft z.B. die Frage, wie die Erweiterung des Planungsraumes auf die 12 Seemeilenzone und die Vorgaben des LROP zu flächenhaften Küstenschutzkonzepten in die jeweiligen RROP übernommen und verankert werden können. Die notwendige Anpassung des bisherigen ingenieur-wissenschaftlichen Ansatzes zur Ermittlung des Sicherheitsstandards ist nicht Gegenstand dieser Dissertation, aber Gegenstand aktueller Forschung. Die in Niedersachsen durch das NDG vorgeschriebenen deterministischen Ansätze (Einzelwertverfahren und Vergleichswertverfahren) müssten durch einen probabilistischen Ansatz ersetzt werden. Hierfür sind aber noch nicht alle notwendigen Bemessungsgrößen verstanden und deren Wirkmechanismen hinreichend bekannt. Dies betrifft insbesondere die Berechnung der Versagenswahrscheinlichkeit der technischen Bauwerke, aber auch die Ermittlung des Schadenspotenzials, die Zusammenführung von Versagenswahrscheinlichkeit und Schadenspotenzial in der Risikogröße sowie deren Begrenzung und Management.

Chapter 8
References

ABPmer. 2005. "Wallasea Island North Bank realignment: environmental statement 2004. Appendix A: review of alternative options considered." Technical Report, R1114, ABPmer.

Abrahamse, J., W. Joenje, and N. van Leeuwen-Seelt. 1976. *Wattenmeer*. Landelijke Vereiniging tot Behoud van de Waddenzee, Harlingen.

Ahlhorn, F. 1997. "Analyse der Nutzungskonflikte und des Managements tidebeeinflusster Strände (mit Dünen) am Beispiel der Nordseeinseln Texel und Norderney". Diploma Thesis, University Oldenburg.

Ahlhorn, F. 2005. "Participation – an overview. Methods and application in Germany. Report of ComCoast project work package 4". Technical Report, RWS-DWW.

Ahlhorn, F., T. Astley-Reid, M. Jones, and W. Snijders. 2006. "From Norcoast to ComCoast. What has happened? Summary of a questionnaire". Technical Report, RWS.

Ahlhorn, F., and T. Klenke. 2006a. "Evaluating European public participation methods for application on a local level in Germany". *New Approach to Harbour, Coastal Risk Management and Education. Proceedings of EUROCOAST–LITTORAL 2006*, 109–117. Gdansk.

Ahlhorn, F., and T. Klenke. 2006b. "Multifunktionale Küstenschutz-Zonen als Instrument im Sinne der Nachhaltigkeit". *TERRAMARE Forschungsberichte*, Volume 15. TERRAMARE und NIHK, 46–51.

Ahlhorn, F., and H. Kunz. 2002a. "The future of historically developed summer dikes and polders: a salt marsh use conflict". *Proceedings of LITTORAL 2002. The Changing Coast*, 365–374. EUROCOAST/EUCC – Porto, Portugal.

Ahlhorn, F., and H. Kunz. 2002b. "Literaturstudie zur Entwicklung und Bedeutung von Sommerdeichen". *Arbeiten aus der Forschungsstelle Küste* 14:99.

Ahlhorn, F., J. Meyerdirks, T. Klenke, T. Astley-Reid, and W. Snijders. 2007a. "Comcoast – identification of sites. Approach to identify feasible sites for the application of multifunctional coastal defence zones". Technical Report, Final, RWS.

Ahlhorn, F., F. Simmering, T. Klenke, J. Meyerdirks, and F. Meyer. 2007b. "GIS-Anwendungen im Rahmen eines nachhaltigen Küstenschutz-Managements". In *Geoinformationen für die Küstenzone. Beiträge des 1. Hamburger Symposiums zur Küstenzone*, edited by K.-P. Traub and J. Kohlus, 59–69. Wichmann, Heidelberg.

Arens, S. 2000. "Vegetationsentwicklung in den Kompensationsflächen der Hauener Hooge 1995–1999. Beweissicherung Küstenschutz Leybucht". Dientsbericht, Forschungsstelle Küste, Norderney.

Arens, S., and E. Götting. 1997. "Ökologische Untersuchungen im deichnahen Salzwiesenbereich des südlichen Jadebusens." Dientsbericht 16/1997, Forschungsstelle Küste, Norderney.

Auhagen, O. 1896. "Zur Kenntnis der Marschwirtschaft. Zwei Abhandlungen". *Landwirtschaftliche Jahrbücher* 25:619–874.

Bakker, J.P., J. Bunje, K. Dijkema, J. Frikke, N. Hecker, B. Kers, P. Körber, J. Kohlus, and M. Stock. 2005. "Salt marshes". In *Wadden Sea Quality Status Report 2004*, edited by K. Essink, C. Dettmann, H. Farke, K. Laursen, G. Lüerßen, H. Marencic and W. Wiersinga, 163–179. CWSS, Wilhelmshaven.

F. Ahlhorn, *Long-Term Perspective in Coastal Zone Development*,
DOI 10.1007/978-3-642-01774-2_8,

Barker, T., et al., [Numerous Authors]. 2007. "Climate change 2007: mitigation. Contribution of working group III to the fourth assessment report of the intergovernmental panel on climate change". Technical Summary, Cambridge University Press, Cambridge, UK.

Bateman, I. 1995. "Environmental and economic appraisal." Edited by T. O'Riordan, In *Environmental Science for Environmental Management*, 45–65 Longman, Harlow.

Behnen, T. 2000. "Der beschleunigte Meeresspiegelanstieg und seine sozio-ökonomischen Folgen: Eine Untersuchung der Ursachen, methodischen Ansätze und Konsequenzen unter besonderer Berücksichtigung Deutschlands". Dissertation, University Hannover.

Behre, K.-E. 1987. "Meeresspiegelschwankungen und Siedlungsgeschichte in den Nordseemarschen". *Vorträge der Oldenburgischen Landschaft* 17:5–47.

Behre, K.-E. 2003. "Eine neue Meeresspiegelkurve für die südliche Nordsee – Transgressionen und Regressionen in den letzten 10.000 Jahren". *Probleme der Küstenforschung* 28:9–63.

Behre, K.-E. 2007. "A new Holocene sea-level curve for the southern North Sea". *Boreas* 36: 82–102.

Belton, V., and T.J. Stewart. 2002. *Multiple Criteria Decision Analysis*. Kluwer Academic Publishers, Boston.

Bijlsma, L. 1994. "Preparing to meet the coastal challenges of the 21st century". *Proceedings of the World Coast Conference in Noordwijk 1993.* Ministry for Transport, Public Works, and Watermanagement, RIKZ.

Blew, J., K. Eskildsen, K. Günther, K. Koffijberg, K. Laursen, P. Potel, H.-U. Rösner, M. van Roomen, and P. Südbeck. 2005. "Migratory birds". In *Wadden Sea Quality Status Report 2004*, edited by K. Essink, C. Dettmann, H. Farke, K. Laursen, G. Lüerßen, H. Marencic and W. Wiersinga, 287–304. CWSS, Wilhelmshaven.

Blindow, H. 1991. "Die Auswirkungen der unterschiedlichen Nutzung der Salzwiesen auf Brutvögel – Ergebnisse des Partnerschaftsprogrammes Dollart-Elisabeth-Außengroden". In *3. Oldenburger Workshop zur Küstenökologie. 20.-22. Feb. 1990*, edited by V. Haeseler and P. Janiesch, 53–55. BIS Universität Oldenburg, Oldenburg.

Blok, D.P. 1984. "Wie alt sind die ältesten niederländischen Deiche? Die Aussagen der frühesten schriftlichen Quellen". *Probleme der Küstenforschung* 15:1–8.

BMU, [Bundesministerium für Umwelt, Naturschutz und Reaktorsicherheit]. 2006. "Integriertes Küstenzonenmanagement in Deutschland. Nationale Strategie für ein integriertes Küstenzonenmanagement." Technical Report, BMU.

Bork, I., and S. Müller-Navarra. 2005. "MUSE – Modellgestützte Untersuchung zu Sturmfluten mit sehr geringen Eintrittswahrscheinlichkeiten. Teilprojekt: Sturmflutsimulation". Technical Report, BSH.

Bosecke, T. 2005. *Vorsorgender Küstenschutz und Integriertes Küstenzonenmanagement (IKZM) an der deutschen Ostseeküste*. Springer Verlag, Berlin Heidelberg.

BR W-E, [Bezirksregierung Weser-Ems]. 1997. "Generalplan Küstenschutz für den Regierungsbezirk Weser-Ems". Technical Report.

BR W-E and NLPV, [Bezirksregierung Weser-Ems, Nationalparkverwaltung Niedersächs-isches Wattenmer]. 2001. "Kompensationsmöglichkeiten für Maßnahmen des Küstenschutzes im Nationalpark "Niedersächsisches Wattenmeer" und angrenzende Bereiche". Technical Report, Bezirksregierung Weser-Ems.

Brahms, A. 1754. *Anfangs-Gründe der Deich- und Wasserbaukunst. 1. und 2. Teil. Nachdruck.* Marschenrat zur Förderung der Forschung im Küstengebiet.

Brandt, K. 1984. "Die mittelalterliche Siedlungsentwicklung in der Marsch von Butjadingen (Landkreis Wesermarsch)". *Siedlungsforschung. Archäologie-Geschichte-Geographie* 2:123–146.

Brandt, K. 1992. "Besiedlungsgeschichte der Nord- und Ostseeküste bis zum Beginn des Deichbaues". In *Historischer Küstenschutz*, edited by J. Kramer and H. Rohde, 17–38. Verlag Konrad Wittwer, Stuttgart.

Brans, J.P., and B. Mareschal. 2005. "PROMETHEE Methods". In *Multiple Criteria Decision Analysis: State of the Art Surveys.*, edited by J. Figuera, S. Greco, M. Ehrgott and C. Henggeler-Antunes, 164–195. Springer Verlag, London.

Brans, J.P., and P. Vincke. 1985. "A preference ranking organisation method". *Management Science* 31(6):647–656.
Brans et al., [J.P. Brans, B. Mareschal and P. Vincke]. 1984. "PROMETHEE: A new family of outranking methods in multi criteria analysis". In *Operational Research '84*, edited by J.P. Brans, 477–490.
Breidert, C. 2005. "Estimation of willingness-to-pay. Theory, measurement and application". Dissertation, University Wien.
Bretschneider, H., K. Lechner, and M. Schmidt. 1993. *Taschenbuch der Wasserwirtschaft*. Verlag Paul Parey, Hamburg.
Breuel, F. 1954. *Geschichte des Anwachsrechts in Ostfriesland*. Vandenhoeck und Ruprecht.
Buchholz, H. 2004. "Raumnutzungs- und Raumplanungsstrategien in den deutschen Meereszonen". *Informationen zur Raumentwicklung* 7:485–490.
Buchwald, K. 1991. *Nordsee. Ein Lebensraum ohne Zukunft?* Verlag Die Werkstatt, Göttingen.
Bungenstock, F. 2006. "Der holozäne Meeresspiegelanstieg südlich der ostfriesischen Insel Langeoog hochfrequente Meeresspiegelbewegungen während der letzten 6000 Jahre". Dissertation, University Bonn.
Bunje, J., and J.L. Ringot. 2003. "Lebensräume im Wandel. Flächenbilanz von Salzwiesen und Dünen im niedersächsischen Wattenmeer zwischen 1966 und 1997 – eine Luftbildauswertung-". Schriftenreihe, Nationalpark Niedersächsisches Wattenmeer, Wilhelmshaven.
Christensen, J.H., T. Carter, and F. Giorgi. 2002. "PRUDENCE employs new methods to assess European climate change". *EOS Trans. (AGU)* 83, no. 13.
Colbourne, L. 2005. "Review of UK literature on public participation and communicating flood risk". Technical Report for the Environment Agency, UK.
Colijn, F., and A. Kannen. 2003. "Projektantrag: Zukunft Küste – Coastal Futures". Technical Report, FTZ Büsum.
Convery, F.J. 2007. "Making a difference – how environmental economists can influence the policy process – a case study of David W. Pearce". *Environmental Resource Economics* 37:7–32.
Cordes, H., D. Mossakowski, and E. Rachor. 1997. "Salzwiesenprojekt Wurster Küste. Abschlussbericht zu den wissenschaftlichen Begleituntersuchungen 1992–1995 im Auftrag der Projekt-Trägergemeinschaft und des Bundesamtes für Naturschutz". Technical Report.
Cox, O. 2005. "Learning from participation. Quick scan of participatory action. Report of ComCoast Project Work Package 4". Technical Report.
CPSL, [Trilateral Working Group on Coastal Protection and Sea Level Rise]. 2001. "Coastal protection and sea level rise. Final report of the trilateral working group on coastal protection and sea level rise". Wadden Sea Ecosystem No. 13, CWSS.
CWSS, [Common Wadden Sea Secretariat]. 1998. "Stade declaration. Trilateral Wadden Sea plan". Technical Report, CWSS.
Daschkeit, A. 2004. "Integriertes Küstenzonenmanagement (IKZM) – sozio-ökologische Perspektiven und Fallstudien". Habilitationsschrift, University Kiel.
Daschkeit, A. 2007. "Integriertes Küstenzonenmanagement (IKZM) als Instrument der räumlichen Planung zur Bewertung von Klimaänderungen im Küstenraum". *Berichte zur deutschen Landeskunde* 2(81):177–187.
de Groot, R.S., M.A. Wilson, and R.M.J. Boumans. 2002. "A typology for the classification, description and valuation of ecosystem functions, goods and services". *Ecological Economics* 41:393–408.
de Ronde, J.G., D. Dilling, and M.E. Phillipart. 1995. "Design criteria along the Dutch coast". *Proceedings of HYDROCOAST '95*, 138–151.
D'Eliso, C. 2007. "Breaching of coastal dikes: preliminary breaching model". LWI Report No. 927 (Floodsite Project), Leichtweiß-Intstitut.
Deutscher Bundestag. 2007a. Grundgesetz für die Bundesrepublik Deutschland.
Deutscher Bundestag. 2007b. Rahmenplan der Gemeinschaftsaufgabe "Verbesserung der Agrarstruktur und des Küstenschutzes" für den Zeitraum 2007 bis 2010. Drucksache 16/5324.
DHV, [Dwars, Heederik and Verhey]. 2005. "ComCoast WP1, Inventarisatie van Locaties. Uitgebreid Plan van Aanpak". Technical Report, RWS.

DHV, [Dwars, Heederik and Verhey]. 2007. "Quick scan climate change adaptation". Technical Report, RIKZ.
Dieckmann, R. 1992. "Chapter morphological structures in German tidal flat areas." *Proceedings of the International Coastal Congress ICC – Kiel '92*, edited by J. Hofstede, H. Sterr and H.-P. Plag, 365–376. Dt. Hochschulschriften.
Dierßen, K. 1987. "Bedeutung, Ziele und Methoden des Naturschutzes für Salzwiesen". In *Salzrasenentwässerung und technische Vorlandgewinnung aus der Sicht des Naturschutzes*, edited by J. Lamp N. Kempf and P. Prokosch, 297–306. WWF-Deutschland, Deutschland.
Dijkema, K.S. 1987. "Geography of salt marshes in Europe". *Z. Geomorph. N.F.* 31:489–499.
Dijkema, K.S., A. Nicolai, J. de Vlaas, C. Smit, H. Jongerius, and H. Nauta. 2001. "Van Landaanwinning naar Kwelderwerken". Technical Report, RWS, Directie Noord Nederland and Alterra, Wageningen.
Dolch, T. 2008. "Spatial high-resolution analysis of morphodynamics and habitat changes in the Wadden Sea". Dissertation, University Kiel.
Doody, P. 2004. "Coastal squeeze – an historical perspective". *Journal of Coastal Conservation* 10(1/2):129–138.
Doody, P., C. Johnston, and B. Smith. 1993. *Directory of the North Sea Coastal Margin*. Joint Nature Conservation Committee.
DVWK, [Deutscher Verband für Wasserwirtschaft und Kulturbau e.V.]. 1985. "Ökono-mische Bewertung von Hochwasserschutzwirkungen – Arbeitsmaterial zum methodischen Vorgehen". *DVWK Mitteilungen* 10.
EA, [Environment Agency]. 2005a. "Building trust with communities". Technical Report.
EA, [Environment Agency]. 2005b. "Work Package 4 – Public Participation. UK Case Studies Schemes and Strategies including costs of participation". unpublished.
EAK, [Advisory Commitee for Coastal Protection]. 1993. "Empfehlungen für die Ausführung von Küstenschutzwerken : EAK 1993 / durch den Ausschuß für Küstenschutz-werke der Deutschen Gesellschaft für Erd- und Grundbau e.V. u. d. Hafenbautechnischen Gesellschaft e.V." *Die Küste* 55:541.
EAK, [Advisory Committee for Coastal Protection]. 2002. "Empfehlungen für die Ausführung von Küstenschutzwerken / durch den Ausschuß für Küstenschutzwerke der Deutschen Gesellschaft für Geotechnik e. V. u. d. Hafenbautechnischen Gesellschaft e. V." *Die Küste* 65:589.
Ebenhöh, W., H. Sterr, and F. Simmering. 1996. "Potentielle Gefährdung und Vulnerabilität der deutschen Nord- und Ostseeküste bei fortschreitendem Klimawandel – Case study based on the common methodology of the IPCC". Technical Report, unpublished.
EC, [European Commission]. 1979. "Council Directive 79/409/EEC of 2 April 1979 on the conservation of wild birds". Birds Directive, European Council.
EC, [European Commission]. 1985. "Council Directive 85/337/EEC of 27 June 1985 on the assessment of the effects of certain public and private projects on the environment". Environmental Impact Assessment, European Parliament and Council.
EC, [European Commission]. 1992. "Council Directive 92/43/EEC of 21 May 1992 on the conservation of natural habitats and of wild fauna and flora". Habitat Directive, European Council.
EC, [European Commission]. 1995. "Communication from the Commission to the Council and the European Parliament on the integrated management of coastal zones". Technical Report.
EC, [European Commission]. 1997. "Council Directive 97/11/EC of 3 March 1997 amending Directive 85/337/EEC on the assessment of the effects of certain public and private projects on the environment". Amended Environmental Impact Assessment, European Parliament and Council.
EC, [European Commission]. 1999. "ESDP – European Spatial Development Perspective. Towards balanced and sustainable development of the territory of the European union". Technical Report.
EC, [European Commission]. 2000. "Directive 2000/60/EC of the European Parliament and of the Council of 23 October 2000 establishing a framework for Community action in the field of water policy". Water Framework Directive, European Parliament and Council.

EC, [European Commission]. 2001. "Directive 2001/42/EC of the European Parliament and of the Council of 27 June 2001 on the assessment of the effects of certain plans and programmes on the environment". Strategic Environmental Assessment, European Parliament and Council.
EC, [European Commission]. 2002. "Recommendations of the European Parliament and Council on the implementation of an integrated coastal zone management in Europe (2002/413/EC), 30 May 2002". Technical Report.
EC, [Euro pean Commission]. 2003a. "Common implementation strategy for the Water Framework Directive (2000/60/EC). Guidance Document No. 8. Public participation in relation to the Water Framework Directive. Produced by Working Group 2.9 – public participation". Technical Report.
EC, [European Commission]. 2003b. "Directive of the European Parliament and Council providing for public participation in respect of the drawing up of certain plans and programmes relating to the environment and amending with regard to public participation and access to justice Council Directive 85/337/EC and 96/61/EC, 26 May 2003". Technical Report.
EC, [European Commission]. 2006. "Proposal for a Directive of the European Parliament and of the Council on the assessment and management of floods". COM(2006) 15 final, SEC(2006) 66, European Parliament and Council.
EEA, [European Environment Agency]. 2001a. "Participatory integrated assessment me-thods: an assessment of their usefulness for the European Environment Agency". Technical Report.
EEA, [European Environment Agency]. 2001b. "Scenarios as tools for integrated environmental assessment". Environmental Issue Report 24, EEA.
EEA, [European Environment Agency]. 2004. "Impacts of Europe's changing climate. An indicator-based assessment". Technical Report 2, EEA.
EEA, [European Environment Agency]. 2005. "Vulnerability and adaptation to climate change in Europe". Technical Report 7, EEA.
EEA, [European Environment Agency]. 2006. "The changing face of Europe's coastal areas". Technical Report 6, EEA.
Ehlers, J. 1988. *The Morphodynamics of the Wadden Sea*. Balkema, Rotterdam.
Eisenführ, F., and M. Weber. 2003. *Rationales Entscheiden*. Springer Verlag, Berlin.
Elsner, A., S. Mai, and C. Zimmermann. 2004. "Risikoanalyse – ein Element des Küstenzonenmanagements". *Coastline Reports* 1:137–147.
Endres, A., and K. Holm-Müller. 1998. *Die Bewertung von Umweltschäden*. Stuttgart.
Engineering Committee for North and Baltic Sea. 1962. "Empfehlungen für den Deichschutz nach der Februar-Strumflut 1962". *Die Küste* 10(1):113–130.
Eppel, D., and K. Ahrendt. 2005. "Wattenmeersedimente: Sedimentinventar Nordfriesisches Wattenmeer". Abschlussbericht, GKSS.
Erbguth, W. 2003. "Wahrung möglicher Belange der Bundesraumordnung in den Ausschließlichen Wirtschaftszonen in Nordsee und Ostsee – Raumordnung in der Ausschließlichen Wirtschaftszone?" Extract of Technical Report.
Erchinger, H.-F. 1970. "Küstenschutz durch Vorlandgewinnung, Deichbau und Deicherhaltung in Ostfriesland". *Die Küste* 19:125–185.
Esser, J. 2001. "Vollständigkeit, Konsistenz und Kompatibilität von Präferenzrelationen". *OR Spektrum* 23:182–201.
Essink, K., C. Dettmann, H. Farke, K. Laursen, G. Lüerßen, H. Marencic, and W. Wiersinga. 2005. *Wadden Sea Quality Status Report 2004*. CWSS, Wilhelmshaven.
EU. 1999. "Eine europäische Strategie für das Küstenzonenmanagement (IKZM): Allgemeine Prinzipien und politische Optionen". Technical Report, EU.
EUROSION. 2004a. "Living with coastal erosion in Europe: sediment and space for sustainability. A guide to coastal erosion management practices in Europe". Technical Report, RIKZ, EUCC, IGN, UAB, BRGM, IFEN and EADS.
EUROSION. 2004b. "Living with coastal erosion in Europe: sediment and space for sustainability. Part II – maps and statistics". Technical Report, RIKZ, EUCC, IGN, UAB, BRGM, IFEN and EADS.

Evans, E., R. Ashley, J. Hall, E. Penning-Rowsell, A. Saul, P. Sayers, C. Thorne, and A. Watkinson. 2004. "Foresight. Future flooding. Scientific summary. Volume I: Future risks and their drivers". Technical Report, Office of Science and Technology, London.

EWT, [Essex Wildlife Trust]. 2004, Winter 2003/2004. The Joan Helliot Visitor Centre at Abbotts Hall Farm. New coastal marshes burst into life. *Newsletter.*

Exo, K.-M. 1994. "Bedeutung des Wattenmeeres für Vögel". In *Warnsignale aus Nordsee und Wattenmeer. Eine aktuelle Umweltbilanz*, edited by J.L. Lozán, E. Rachor, K. Reise, H.v. Westernhagen and W. Lenz, 261–270. Wissenschaftliche Auswertung, Hamburg.

Feilbach, M. 2004. "Entwurf eines Integrierten Küstenzonenmanagementplans für die Odermündung". IKZM-Oder Bericht 2, Institut für Geographie, University Greifswald.

Flinterman, M., and A. Glasius-Meier. 2005. "The citizen's perspective in a socio-economic evaluation? A quick scan of the current state of affairs. Report of Work Package 2 of the ComCoast project". Technical Report.

Flügel, R., and A. Dziatzko. 2008. "Multifunktionale Küstenschutz-Zonen in Niedersachsen. Untersuchung der Umsetzungsmöglihckeiten aus Sicht der Raumplanung". Diploma Thesis, Technical University Dortmund.

Führböter, A. 1987. "Über den Sicherheitszuwachs im Küstenschutz durch eine zweite Deichlinie". *Die Küste* 45:181–208.

Gätje, C., and K. Reise. 1998. *Ökosystem Wattenmeer. Austausch-, Transport- und Stoffumwandlungsprozesse.*

Gee, K., A. Kannen, and K. Licht-Eggert. 2006. "Raumordnerische Bestandsaufnahme für die deutsche Küsten- und Meeresbereiche". Berichte Nr. 38, FTZ Büsum.

Girod, B. 2006. "Why six baseline scenarios? A research on the reason for the growing baseline uncertainty of the IPCC scenarios". Diploma Thesis, Technical University Zurich.

Giszas, H. 2003. "Sturmflutschutz: Herausforderungen und Sicherheitskonzept". In *Proceedings of HTG – Flüsse, Kanäle, Häfen, Sept. 2003*, 241–258. HTG – Hafenbautechnische Gesellschaft.

Götting, E., W. Heiber, and K. Heljen. 2002. "Beweissicherung Küstenschutz Leybucht. Entwicklung der terrestrischen Wirbellosenfauna in ausgedeichten und volleingedeichten Bereichen des ehemaligen Sommerpolders Hauener Hooge (1995–1999)". Bericht 5/2002, Forschungsstelle Küste, Norderney.

Gouldby, B., and P. Samuels. 2005. "Language of risk. Project definitions." Technical Report of the project FLOOD*site* – Integrated Flood Risk Analysis and Management Methodologies, T32-04-01, HR Wallingford.

Greiving, S. 2002. *Räumliche Planung und Risiko.* Gerling Akademie Verlag, München.

Grossmann, I., K. Woth, and H. von Storch. 2007. "Localization of global climate change: storm surge scenarios for Hamburg in 2030 and 2085". *Die Küste* 71:169–182.

Grunwald, A. 2002. *Technikfolgenabschätzung – eine Einführung.* Ed. Sigma.

Guitouni, A., and J.-M. Martel. 1998. "Tentative guidelines to help choosing an appropriate MCDA method". *European Journal of Operational Research* 109:501–521.

Guzman, R.M., and C.D. Kolstad. 2007. "Researching preferences, valuation and hypothetical biases". *Environmental Resource Economics* 37:465–487.

Halcrow. 1998a. "Essex sea wall management. Executive summary". Technical Report, Environment Agency Anglian Region.

Halcrow. 1998b. "Essex sea wall management. Strategy report". Technical Report, Environment Agency Anglian Region.

Hampicke, U. 1991. "Neoklassik und Zeitpräferenz: der Diskontierungsnebel". Edited by F. Beckenbach, *Die ökologische Herausforderung für die ökonomische Theorie.* Marburg, 127–150.

Hampicke, U. 1992. *Ökologische Ökonomie. Individuum und Natur in der Neoklassik.* Opladen.

Hansen, J., M. Sato, R. Ruedy, K. Lo, D.W. Lea, and M. Medina-Elizade. 2006. "Global temperature change". *PNAS* 103 (39): 14288–14293.

Hansjürgens, B. 2004. "Economic valuation through cost-benefit analysis – possibilities and limitations". *Toxicology* 205(3):241–252.

Hartje, V., I. Meyer, and J. Meyerhoff. 2002. Kosten einer möglichen Klimaänderung auf Sylt. In *Klimafolgen für Mensch und Küste: Am Beispiel der Nordseeinsel Sylt*, edited by A. Daschkeit and P. Schottes, 181–218. Springer, Berlin.

Hartung, W. 1983. "Die Leybucht (Ostfriesland) – Probleme ihrer Erhaltung als Naturschutzgebiet". *Neues Archiv für Niedersachsen* 32(4):355–387.

Henocque, Y. 2003. "Development of process indicators for coastal zone management assessment in France". *Ocean and Coastal Management* 46:363–379.

Heydemann, B. 1987. "Bedeutung, Ziele und Methoden des Naturschutzes für Salzwiesen". In *Salzwiesen: Geformt von Küstenschutz, Landwirtschaft oder Natur? Intern. Fachtagung zu Perspektiven für Schutz und Pflege von Salzwiesen im Wattenmeer*, edited by J. Lamp, N. Kempf and P. Prokosch, 71–82. WWF-Deutschland, Deutschland.

Heydemann, B., and J. Müller-Karch. 1980. *Biologischer Atlas Schleswig-Holstein. Lebensgemeinschaften des Landes*. Wachholtz, Neumünster.

Hillen, R., and T. de Haan. 1993. "Development and implementation of the coastal defence policy for The Netherlands". In *Coastline of the Southern North Sea*, edited by R. Hillen and H.J. Verhagen, ASCE, Louisiana, 188–201.

Hillen, R., L. Bijlsma, and R. Misdorp. 1992. "Towards sustainable development of coastal zones. Report on the activities of the coastal zone management subgroup of IPCC." Edited by J. Hofstede, H. Sterr and H.-P. Plag, In *Proceedings of the International Coastal Congress, ICC-Kiel'92*, P. Lang, Frankfurt am Main, 442–453.

Hofmeister, A.E. 1984. "Zum mittelalterlichen Deichbau in den Elbmarschen bei Stade". *Probleme der Küstenforschung* 15:41–50.

Hofstede, J.L.A. 1999. "Process-response analysis for Hörnum tidal inlet in the German sector of the Wadden Sea". *Journal of Quaternary International* 60:107–117.

Hofstede, J.L.A., and R. Schirmacher. 1996. "Volandmanagement in Schleswig-Holstein". *Die Küste* 58:61–73.

Holman, I.P., R.J. Nicholls, P.M. Berry, P.A. Harrison, E. Audsley, S. Shackley, and M.D.A. Roundsevell. 2005a. "A regional multi-sectoral and integrated assessment of the impacts of climate change and socio-economic change in the UK. Part II: results". *Climatic Change* 71:43–73.

Holman, I.P., M.D.A. Roundsevell, S. Shackley, P.A. Harrison, R.J. Nicholls, P.M. Berry, and E. Audsley. 2005b. "A regional multi-sectoral and integrated assessment of the impacts of climate change and socio-economic change in the UK. Part I: Methodology". *Climatic Change* 71:9–41.

Homeier, H. 1974. "Untersuchungen zum Verlandungsfortschritt im Bereich der Leybucht". In *Jahresbericht 25/1973*, 11–32. Forschungsstelle Norderney.

Houtekamer, N., T. Astley-Reid, T. Stroobandt, I. Bliss, and F. Ahlhorn. 2007. "Public participation: why you aren't doing it. Tips and solutions from ComCoast pilot sites". Final Report of Work Package 4.

Hübler, M., G. Klepper, and S. Peterson. 2007. "Costs of climate change. The effects of rising temperatures on health and productivity". Working Paper 1321, Kiel Institute for the World Economy.

Hulme, M., J. Turnpenny, and G. Jenkins. 2002. "Climate change scenarios for the United Kingdom. The UKCIP02 briefing report." Technical Report, UK Climate Impacts Programme (UKCIP, Tyndall Centre for Climate Change Research and Hadley Centre for Climate Prediction and Research).

IAU, [Impact Assessment Unit of the Oxford Brooks University]. 2003. "Review of international legal instruments, policies and management in respect of the Wadden Sea region". Technical Report, Wadden Sea Forum.

ICLC, [Imperial College London Consultants]. 2005. "The relationship between the EIA and SEA Directives". Technical Report, European Commission.

INFRAM. 2007. "Study on effectiveness of ComCoast solutions". Technical Report, INFRAM, Marknesse (NL).

Ingenieur-Kommission Niedersachsen. 1962. "Die Sturmflut vom 16./17. Februar 1962 im niedersächsischen Küstengebiet". *Die Küste* 10(1):17–54.

IPCC, [Intergovernmental Panel on Climate Change]. 1990. *Climate Change, the IPCC Scientific Assessment*. WMO/UNEP.

IPCC, [Intergovernmental Panel on Climate Change]. 1996. *Climate Change 1995: Impacts Adaption and Mitigation of Climate Change: Scientific-Technical Analysis*. Edited by R. H. Moss, R. T. Watson, M. C. Zinyowera and D. J. Dokken. Cambridge University Press, Cambridge, UK.

IPCC, [Intergovernmental Panel on Climate Change]. 2000. *Special Report on Emissions Scenarios*. WMO/UNEP.

IPCC, [Intergovernmental Panel on Climate Change]. 2007. "Climate Change 2007: Synthesis Report". Summary for Policymakers, IPCC.

Janssen, T. 1992. *Die Leybucht: Natur- und Küstenschutz. Eine Dokumentation*. SKN, Norden.

Jelgersma, S. 1979. "Sea-level changes in the North Sea basin". In *The Quaternary History of the North Sea*, edited by R.T.E. Schüttenhelm, E. Oele and J.A. Wiggers, 233–248. Acta Univ. Ups. Symp. Univ. Annum Quingentesium Celebrantis 2, Uppsala.

Jensen, J., and C. Mudersbach. 2005. "MUSE – Modellgestützte Untersuchung zu Sturmfluten mit sehr geringen Eintrittswahrscheinlichkeiten. Teilprojekt: Statistisch-probabilistische Extremwertanalyse". Technical Report, University Siegen.

Jeschke, A. 2004. "Raumplanung als vorsorgendes Instrument im Küstenschutzmanagement". Master Thesis, University Oldenburg.

Jonas, M., T. Staeger, and C.-D. Schönwiese. 2005. "Berechnung der Wahrscheinlichkeiten für das Eintreten von Extremereignissen durch Klimaänderungen. Schwerpunkt Deutschland". Technical Report 7, Umweltbundesamt (Ed.).

Kaiser, G. 2006. "Risk and vulnerability analysis to coastal hazards – an approach to integrated assessment". Dissertation, University Kiel.

Kaiser, G., S. Reese, H. Sterr, and H.-J. Markau. 2004. "ComRisk – Common strategies to reduce the risk of storm floods in coastal lowlands. Subproject 3: Public perception of coastal flood defence and participation in coastal flood defence planning". Technical Report.

Kaul, J.-A., and C. Reins. 2000. "Abschlussbericht der Sensitivitätsanalyse zu einem integrierten Küstenschutzkonzept für die Küstenniederung "Timmendorfer Strand/Scharbeutz"". Technical Report.

Kiese, M., and B. Leineweber. 2001. "Risiko einer Küstenregion bei Klimaänderung. Ökonomische Bewertung und räumliche Modellierung des Schadenspotentials in der Unterweserregion". Technical Report 25.

Kinder, M., I. Vagts, and W. Züghart. 1993. "Salzwiesenprojekt Wurster Küste. Zwischenbericht über die wissenschaftlichen Begleituntersuchungen 1992. Floristischer und vegetationskundlicher Teil". Technical Report.

Klaus, J. 1986. "Vorteilsanalyse zum Fachplan Küstenschutz Sylt". Technical Report, Ministry of Rural Areas, Agriculture, Food and Tourism Schleswig-Holstein.

Klaus, J., and R.F. Schmidtke. 1990. "Bewertungsgutachten für Deichbauvorhaben an der Festlandsküste –Modellgebiet Wesermarsch–". Technical Report.

Klein, R.J.T., and R.J. Nicholls. 1999. "Assessment of coastal vulnerability to climate change". *Ambio* 28(2):182–187.

Klenke, T., F. Ahlhorn, and A. Jeschke. 2006. "Multifunktionale Küstenschutzräume als Baustein eines integrierten Küstenzonenmanagements in Nieder-sachsen". *Wasser und Abfall* 8(9):15–19.

Klug, H., M. Hamann, S. Reese, and T. Rohr. 1998. "Wertermittlung für die potentiell sturmflutgefärdeten Gebiete an den Küsten Schleswig-Holsteins". Technical Report for Ministry of Rural Areas, Agriculture, Food and Tourism Schleswig-Holstein.

Koffijberg, K., J. Blew, K. Eskildsen, K. Günther, B. Koks, K. Laursen, L.-M. Rasmussen, P. Südbeck, and P. Potel. 2003. "High tide roosts in the Wadden Sea: A review of bird distribution, protection regimes and potential sources of anthropogenic disturbance. A report of the

Wadden Sea Plan Project 34". *Wadden Sea Ecosystems* 16. Common Wadden Sea Secretariat, Wilhelmshaven, 120.

Koffijberg, K., L. Dijksen, B. Hälterlein, K. Laursen, P. Potel, and P. Südbeck. 2005. "Breeding birds". In *Wadden Sea Quality Status Report 2004*, edited by K. Essink, C. Dettmann, H. Farke, K. Laursen, G. Lüerßen, H. Marencic and W. Wiersinga, 275–286. CWSS, Wilhelmshaven.

Kortenhaus, A. 2003. "Probabilistische Bemessung von Seedeichen". Dissertation, Technical University Braunschweig.

Kortenhaus, A., P. Fröhle, J. Jensen, N. von Lieberman, S. Mai, C. Miller, K. Peters, and H. Schüttrumpf. 2007. "Probabilistische Bemessung von Bauwerken". *HANSA* 144(4):68–76.

Koziar, C., and V. Renner. 2005. "MUSE – Modellgestützte Untersuchung zu Sturmfluten mit sehr geringen Eintrittswahrscheinlichkeiten. Teilprojekt: Numerische Berechnung physikalisch konsistenter Wetterlagen mit Atmosphärenmodellen". Technical Report, DWD.

Kramer, J. 1989. *Kein Deich, kein Land, kein Leben – Geschichte des Küstenschutzes an der Nordsee*. Verlag Gerhard Rautenberg, Leer.

Kramer, J. 1992. "Entwicklung der Deichbautechnik an der Nordseeküste". In *Historischer Küstenschutz*, edited by J. Kramer and H. Rohde, 63–109. Verlag Konrad Wittwer, Stuttgart.

Kramer, J., and H. Rohde. 1992. *Historischer Küstenschutz. Deichbau, Inselschutz und Binnenentwässerung and Nord- und Ostsee*. Verlag Konrad Wittwer, Stuttgart.

Krämer, R. 1984. "Landesausbau und mittelalterlicher Deichbau in der hohen Marsch von Butjadingen. Ergebnisse historisch-geographischer Untersuchungen". *Siedlungsforschung. Archäologie-Geschichte-Geographie* 2:147–164.

Kristensen, P. 2003. "EEA Core Set Indicators". Technical Report, EEA.

Kunz, H. 1991. "Klimaänderungen, Meerespiegelanstieg, Auswirkungen auf die nieder-sächsische Küste". *Mitteilungen des Franzius-Instituts* 72:323–351.

Kunz, H. 1994. "Die Einwirkungen des Meeres und des Menschen auf das Küstengebiet - Küstenschutz und Ökologie: Ein Widerspruch? – Aufgaben und Strategien". *Mitteilungen des Franzius-Instituts* 75:9–51.

Kunz, H. 1999a. "Groyne field technique against the erosion of salt marshes – renaissance of a soft engineering approach". In *Proceedings Fourth International Conference on the Mediterranean Coastal Environment (MEDCOAST)*, 1477–1490. MEDCOAST.

Kunz, H. 1999b. "The Leyhörn Project – an integrated engineering response to multifarious demands of a coastal community". In *Proc. Fifth Intern. Conf. on Coastal and Port Engineering in Developing Countries*, Volume 3, 2176–2186. COPEDEC.

Kunz, H. 2004a. "Küstenschutz- und Küstenzonenmanagement: Gesamtschau einer Dokumentation". In *Küstenschutz- und Küstenzonenmanagement – eine Dokumentation*, Arbeiten aus der Forschungsstelle Küste, Heft 15, 13–78. Niedersächsisches Landesamt für Ökologie.

Kunz, H. 2004b. "Sicherheitsphilosophie für den Küstenschutz". *Jahrbuch der Hafenbautechnischen Gesellschaft* 54:253–287.

Lampe, E. 2008. "Risiko-PR der Deichverbände". Diploma-Thesis, in preparation, University Bonn, pp. 180.

Lange, H., M. Haarmann, A. Wiesner-Steiner, and E. Voosen. 2005. "Politisch-administrative Steuerungsprozesse (PAS). Endbericht des Teilprojektes 4 im Verbundvorhaben Klimawandel und präventives Risiko- und Küstenschutzmanagement an der deutschen Nordseeküste (KRIM)". Technical Report.

Lange, H., A. Wiesner, M. Haarmann, and E. Voosen. 2007. "Handeln nur auf Basis sicheren Wissens. Die Konstruktion des Risikos aus Sturmfluten und Klimawandel im politisch-administrativen System". In *Land unter? Klimawandel, Küstenschutz und Risikomanagement in Nordwestdeutschland: die Perspektive 2050*, 145–166. Oekom Verlag.

LAWA, [Länderarbeitsgemeinschaft Wasser]. 1981. "Grundzüge der Nutzen-Kosten-Untersuchungen". Technical Report.

LKFriesland, [Landkreis Friesland]. 2002. "Kleigewinnung für die Verstärkung des Elisabethgrodendeiches in der Gemeinde Wangerland – Raumordnerische Beurteilung". Technical Report, III. Oldenb. Deichband and NLWK.

Lorenzen, J.M. 1955. "Hundert Jahre Küstenschutz an der Nordsee". *Die Küste* 3(1/2):18–32.
Lorenzen, J.M. 1966. "Über Aufgaben und Organisation des Küstenausschusses Nord- und Ostsee". *Die Küste* 14(2):1–4.
Louisse, C.J., and F. van der Meulen. 1991. "Future coastal defence in The Netherlands: strategies for protection and sustainable development". *Journal of Coastal Research* 7(4):1027–1041.
Lozán, J.L., E. Rachor, K. Reise, H. von Westernhagen, and W. Lenz. 1994. *Warnsignale aus dem Wattenmeer*. Blackwell, Berlin.
LSBG, [Landesbetrieb für Straßen, Brücken und Gewässer]. 2007. "Hochwasserschutz in Hamburg. Bauprogramm 2007". Technical Report, LSBG.
Lüders, K., and G. Leis. 1964. *Niedersächsisches Deichgesetz – Kommentar*. Verlag Wasser und Boden, Hamburg.
Mai, S. 2004. "Klimafolgenanalyse und Risiko für eine Küstenzone am Beispiel der Jade-Weser Region". Dissertation, University Hannover.
Mai, S., and A. Bartholomä. 2000. "The missing mud flats of the Wadden Sea: a reconstruction of sediments and accomodation space lost in the wake land reclamation". Edited by M.T. Delafontaine, B.W. Flemming and G. Liebezeit, *Muddy Coast Dynamics and Resource Management*. 257–272. Elsevier, Amsterdam.
Markau, H.-J. 2003. "Risikobetrachtung von Naturgefahren. Analyse, Bewertung und Management des Risikos von Naturgefahren am Beispiel der sturmflugefährdeten Küstenniederungen Schleswig-Holsteins". Dissertation, University Kiel.
Matulla, C., W. Schöner, H. Alexandersson, H. von Storch, and X.L. Wang. 2007. "European storminess: late 19th century to present". *Climate Dynamics*, no. DOI 10.1007/s00382-007-0333-y:6.
MELF, [Niedersächsisches Ministerium für Ernährung, Landwirtschaft und Forsten]. 1973. "Generalplan Küstenschutz Niedersachsen". Technical Report.
MELFF, [Arbeitsgruppe des Ministeriums für Ernährung, ländliche Räume, Forsten u. Fischerei (MELFF) und des Minsteriums für Natur und Umwelt (MNU) des Landes Schleswig-Holstein]. 1995. "Vorlandmanagement in Schleswig-Holstein". Technical Report, MELFF – MNU.
Meyer, C., and G. Ragutzki. 1999. "KFKI Forschungsvorhaben Sedimentverteilung als Indikator für morphodynamische Prozesse". Dienstbericht 21/1999, Forschungsstelle Küste, Norderney.
Meyer, R. 1926. *Die rechtliche Struktur und die wirtschaftlichen Verhältnisse der Deichbände im Oldenburgischen Staatsgebiet*. Ad. Littmann, Oldenburg.
Meyerdirks, J. 2008. "Analyse der Klimasensitivität von Gebieten mit besonderer Bedeutung für Natur und Landschaft im Bereich der deutschen Nordseeküste". Dissertation, University Bremen.
Meyerdirks, J., and F. Ahlhorn. 2007a. "Case Study 6 – Nessmersiel. The participatory integrated assessment process". Technical Report, ICBM, University of Oldenburg.
Meyerdirks, J., and F. Ahlhorn. 2007b. "ComCoast Nessmersiel – Unterlagen zur Vorbereitung des Workshops am 24. April 2007". Technical Report, ICBM, University of Oldenburg.
Meyerdirks, J., and F. Ahlhorn. 2008. "Combining integrated assessment and progressive stakeholder engagement to support sustainable land use management in coastal zones". *Proceedings of EUROCOAST–LITTORAL 2008*, Venice. 8.
Michaelis, H. 1968. "Biologische-sedimentologische Untersuchungen des Wurster Watts von Spieka bis Neufeld". In *Jahresbericht XVIII/1966*, 71–82. Forschungsstelle Küste, Norderney.
ML, [Niedersächsisches Ministerium für den ländlichen Raum, Ernährung, Landwirtschaft und Verbraucherschutz]. 2005. "Umweltbericht für die Novellierung des Landesraumordnungsprogramms (Auszug)". Technical Report.
ML, [Niedersächsisches Ministerium für den ländlichen Raum, Ernährung, Landwirtschaft und Verbraucherschutz]. 2008. "Landesraumordnungsprogramm des Landes Niedersachsen". Latest Amendment, 30th January 2008, ML.
MLR, [Ministerium für ländliche Räume, Landesplanung, Landwirtschaft und Tourismus des Landes Schleswig-Holstein]. 2001. "Generalplan Küstenschutz. Integriertes Küstenschutzmanagement in Schleswig-Holstein". Technical Report.

ML-RVOL, [Ministerium für den ländlichen Raum, Ernährung, Landwirtschaft und Verbraucherschutz – Regierungsvertretung Oldenburg]. 2005. "Raumordnerisches Konzept für das niedersächsische Küstenmeer". Technical Report, Regierungsvertretung Oldenburg.

MLUR, [Ministerium für ländliche Räume, Landesplanung, Landwirtschaft und Tourismus]. 2001. "Vorlandmanagementkonzept. Erfahrungsbericht 1995–2000". Technical Report, MLUR.

MPI-M. 2008. Pictures free download from http://www.mpimet.mpg.de/, last visit May 2008.

MU, [Ministerium für Umwelt]. 2004. "Neubekanntmachung des Niedersächsischen Deichgesetzes – Amendment of the State Law on Dikes for Lower Saxony". Nds. GVBl. 6/2004.

MU, [Ministerium für Umwelt]. 2006. "Entwicklung der 10 Grundsätze für einen effektiveren Küstenschutz". Technical Report.

Munda, G. 2005. "Multiple criteria decision analysis and sustainable development". In *Multiple Criteria Decision Analysis: State of the Art Surveys*, edited by J. Figuera, S. Greco, M. Ehrgott and C. Henggeler-Antunes, 953–986. Springer Verlag, Berlin.

Munton, R. 2003. "Deliberative democracy and environmental decision-making". In *Negotiating environmental change. New Perspectives from social science*, edited by F. Berkhout, M. Leach and I. Scoones, 109–136. Chelten-ham, UK.

MVenW, [Ministry of Transport, Public Works and Water Management]. 1990. "A new coastal defence policy for the Netherlands". Technical Report.

MVenW, [Ministry of Transport, Public Works and Water Management]. 2002. "Naar integraal kustzonebeleid. Beleidsagenda voor de kust". Technical Report.

MVenW, [Ministry of Transport, Public Works and Water Management]. 2005a. "Floris study – full Report". DWW-2006-014.

MVenW, [Ministry of Transport, Public Works and Water Management]. 2005b. "Veiligheid Nederland in Kaart. Hoofdrapport onderzoek overstromingsrisico's". DWW-2005-081.

MVenW, [Ministry of Transport, Public Works and Water Management]. 2005c. "Veiligheid Nederland in Kaart. Tussenstand onderzoek overstromingsrisico's". DWW-2005-074.

NDG. 1963. "Niedersächsisches Deichgesetz – state law on dikes for Lower Saxony". Nds. GVBl. 6/2004.

Niemeyer, H.-D. 1986. "Ausbreitung und Dämpfung des Seegangs im See- und Wattengebiet von Norderney". *Jahresbericht der Forschungsstelle Küste* 37:49–95.

Niemeyer, H.-D., and R. Kaiser. 1999. "Mittlere Tidewasserstände". In *Umweltatlas Wattenmeer – Band 2. Wattenmeer zwischen Elb- und Emsmündung*, edited by Nationalparkverwaltung Niedersächsisches Wattenmeer and Umweltbundesamt, 24–25. Ulmer, Stuttgart.

NLÖ, [Niedersächsisches Landesamt für Ökolgie]. 2000. "Projektgruppe zur Verbesserung des Verfahrensmanagement im Küstenschutz". Technical Report, NLÖ.

NLP-V and UBA, [Nationalparkverwaltung Niedersächsisches Wattenmeer and Umweltbundesamt]. 1999. *Umweltatlas Wattenmeer. Band 2. Wattenmeer zwischen Elb- und Emsmündung*. Ulmer, Stuttgart.

NLWK Norden, [Niedersächsischer Landesbetrieb für Wasserwirtschaft und Küstenschutz – Betriebsstelle Norden]. 2003. "Vorlandmanagementplan für den Bereich der Deichacht Norden". Technical Report.

NLWKN, [Niedersächsischer Landesbetrieb für Wasserwirtschaft, Küsten- und Naturschutz]. 2007a. "Fact-Sheet about the Pilot Area Nessmersiel." Flyer for the WP3 of the ComCoast project, RWS.

NLWKN, [Niedersächsischer Landesbetrieb für Wasserwirtschaft, Küsten- und Naturschutz]. 2007b. "Generalplan Küstenschutz Niedersachsen/Bremen – Festland". Technical Report.

NNatG. 2007. "Niedersächsisches Naturschutzgesetz". In der Fassung der Bekanntmachung vom 11. April 1994 (Nds. GVBl. S. 155), zuletzt geändert durch Artikel 4 des Gesetzes vom 26. April 2007 (Nds. GVBl. S. 161).

NORCOAST Project Secretariat. 2000. "NORCOAST – recommendations on improved integrated coastal zone management in the North Sea region". Technical Report, County of North Jutland.

NPG-SH. 1996. "Gesetz zum Schutz des schleswig-holsteinischen Wattenmeeres". Vom 22. Juli 1985 (GVOBl. Schl.-H. S. 202 zuletzt ersetzt durch Verordnung vom 24. Oktober 1996 (GVOBl. S. 652).

NROG. 2007. "Niedersächsisches Gesetz über Raumordnung und Landesplanung". Nds. GVBl. Nr. 17/2007 S.223.

NUVPG. 2007. "Niedersächsisches Gesetz über die Umweltverträglichkeitsprüfung in der Fassung vom 30. April 2007". Nds. GVBl. Nr. 13/2007.

NWattNPG. 2001. "Gesetz über den Nationalpark "Niedersächsisches Wattenmeer"". Vom 11. Juli 2001, Nds. GVBl. 2001/443.

NWG. 2004. "Niedersächsisches Wassergesetz". Last change 2007.

OECD. 2001. "Citizens as partners. OECD handbook on information, consultation and public participation in policy-making". Technical Report.

Oedekerk, M. 2006. "Van dijkversterking naar dijkvervaging. Onderzoek naar de moge-lijkheiden van brede waterkeringen in Groningen". Master Thesis, University Groningen.

Olsen, S.B. 2003. "Framework and indicators for assessing progress in integrated coastal zone management initiatives". *Ocean and Coastal Management* 46:347–361.

Oost, A.P. 1995. "Dynamics and sedimentary development of the Dutch Wadden Sea with emphasis on the Frisian inlet". Geologica Ultaiectina, Mededelingen van Faculteit Aardwetenschappen, Universiteit Utrecht.

OSPAR, [OSPAR Commission]. 2000. "Quality Status Report 2000. Region II Greater North Sea". Technical Report, OSPAR.

Otto, H.-J. 2004. "Sicherheitsstandards im Hamburger Küstenschutz". *Jahrbuch der Hafenbautechnischen Gesellschaft* 54:47–51.

Oumeraci, H., and A. Kortenhaus. 2002. "Risk-based design of coastal flood defences: a suggestion for a conceptual framework". *Proceedings ICCE 2002* 2:2399–2411.

Pahl-Wostl, C., C.C. Jaeger, S. Rayner, C. Schäfer, M.B. von Asselt, D.M. Imboden, and A. Vckovski. 1998. "Regional integrated assessment and the problem of indeterminacy". In *View from the Alps. Towards Regional Assessment of Climate Change*, edited by P. Cebon, U. Dahinden, H.C. Davies, D.M. Imboden, C.C. Jaeger, 435–497. MIT Press, Cambridge.

Parmesan, C., and G. Yohe. 2003. "A globally coherent fingerprint of climate change impacts across natural systems". *Nature* 421:37–42.

Parry, M.L. 2000. "Assessment of potential effects and adaptations for climate change impacts across natural systems". *Nature* 421:37–42.

Parry et al., [Numerous Authors]. 2007. "Technical Summary". In *Climate Change 2007: Impacts, Adaptation and Vulnerability. Contribution of Working Group II to the Fourth Assessment Report of the Intergovernmental Panel on Climate Change*, 23–78. Cambridge University Press, Cambridge, UK.

Patel, M., K. Kok and D.S. Rothman. 2007. "Participatory scenario construction in land use analysis: an insight into the experiences created by stakeholder involvement in the Northern Mediterranean". *Land Use Policy* 24(3):546–561.

Pearce, D.W., and R.K. Turner. 1990. *Economics of Natural Resources and the Environment*. Harvester Wheatsheaf, London.

Peerbolte, E.B., J.J. de Ronde, L.P.M. de Vrees, M. Mann, and G. Baarse. 1991. *Impacts of Sea Level Rise on Dutch Society, a Description of the ISOS (=Impact of Sea Level Rise on Society) Case Study for the Netherlands*. Delft Hydraulics and Rijkswatertstaat, Delft and The Hague, The Netherlands.

Peters, H.P., and H. Heinrichs. 2007. "Das öffentliche Konstrukt der Risiken durch Sturmfluten und Klimawandel". In *Land unter? Klimawandel, Küstenschutz und Risikomanagement in Nordwestdeutschland: die Perspektive 2050*, 115–144. Oekom Verlag.

Peters, K.-H. 1992. "Die Entwicklung des Deich- und Wasserrechts". In *Historischer Küstenschutz*, edited by J. Kramer and H. Rohde, 183–206. Verlag Konrad Wittwer, Stuttgart.

Petersen, J., and E.J. Lammerts. 2005. "Dunes". In *Wadden Sea Quality Status Report 2004*, edited by K. Essink, C. Dettmann, H. Farke, K. Laursen, G. Lüerßen, H. Marencic and W. Wiersinga, 241–258. CWSS, Wilhelmshaven.

Petersen, K. 1998. "Küstenschutz und Naturschutz im Zielkonflikt – ist der Küstenschutz am Ende?" *Wasser und Boden* 8:45–48.

Petersen, M. 1966. "Die zweite Deichlinie im Schutzsystem der deutschen Nordseeküste". *Die Küste* 14(2):101–106.

Pickaver, A., C. Gilbert, and F. Breton. 2004. "An indicator set to measure the progress in the implementation of integrated coastal zone management in Europe". *Ocean and Coastal Management* 47:449–462.

Plapp, S.T. 2001. "Perception and evaluation of natural risks – interim report on first results of a survey in six districts in Germany". Working Paper 1, University Karlsruhe.

Plapp, S.T. 2003. "Wahrnehmung von Risiken aus Naturkatastrophen. Eine empirische Untersuchung in sechs gefährdeten Gebieten Süd- und Westdeutschland". Dissertation, University Karlsruhe.

Post, J.C., and C.G. Lundin. 1996. "Guidelines for integrated coastal zone management". ESD Studies and Monographs, No. 9, The World Bank.

Prange, W. 1986. "Die Bedeichungsgeschichte der Marschen in Schleswig-Holstein". *Probleme der Küstenforschung* 16:1–53.

Provincie Noord-Holland. 2005. "Zwakke Schakels Noord-Holland". Technical Report.

Ravensdale, J.R. 1981. "A comparative note on the exploitation and draining of the peat fens near the Wash". In *Proceedings of the symposium on peat lands below sea level*, 74–84. International Institute for Land Reclamation and Improvement/ILRI, Wageningen.

Reese, S., H.-J. Markau, and H. Sterr. 2002. "MERK – Mikroskalige Evaluation der Risiken in überflutungsgefährdeten Küstenniederungen". Technical Report, Universiy Kiel, FTZ Westküste.

Reineck, H.-E. 1978. "Die Watten der deutschen Nordseeküste". *Die Küste* 32:64–81.

Renn, O., R. Carius, H. Kastenholz, and M. Schulze. 2005. "ERiK – Entwicklung eines mehrstufigen Verfahrens der Risikokommunikation". Technical Report.

Renn, O., and T. Webler. 1994. "Konfliktbewältigung durch Kooperation in der Umweltpolitik – Theoretische Grundlagen und Handlungsvorschläge". In *Umweltökonomischen Studenteniniative OIKOS an der Hochschule St. Gallen*, edited by Kooperationen für die Umwelt. Im Dialog zum Handeln. Zürich (Ruegger Verlag: 1994), 11–52.

Richwien, W., and H.-D. Niemeyer. 2007. "INTBEM – Integrierte Bemessung von See- und Ästuardeichen". Technical Report, University Essen and NLWKN – Forschungsstelle Küste.

Ridder, D., and C. Pahl-Wostl. 2005. "Participatory integrated assessment in local level planning". *Regional Environmental Change* 5:188–196.

RIKZ, [Rijksinstituut voor Kust en Zee]. 2002. "Waterbewustzijn in Nederland: Leren van risicobewustwordingsprocessen in het buitenland". RIKZ/2003.005, RIKZ, BWD and ERGO.

RIKZ, [Rijksinstituut voor Kust en Zee]. 2006. "Kwetsbaarheid voor overstroming van dijkring Centraal Holland". Technical Report, 2006.108w, RIKZ.

Risk Commission. 2003. "Ad hoc Commssion on "Revision of risk analysis procedures and structures as well as of standard setting in the field of environmental health in the Federal Republic of Germany"." Technical Report.

Roose, K. 2006. "Innovo et emergo. Procesaabeveling voor een multifunctionele kustzone in Zeeland". Master Thesis, Technical University Eindhoven.

Roy, B. 1985. "Méthodologie multicritère d'aide à la décision". *Economica* XXII:423.

Roy, B. 1996. *Multi Criteria Methodology for Decision Aiding*. Springer Verlag, Berlin.

Royal Haskoning. 2007. "Integrated coastal zone development. A process approach based on ComCoast experiences". Technical Report of WP1 of the ComCoast project, RWS, Delft.

Ruhland, A. 2004. "Entscheidungsunterstützung zur Auswahl von Verfahren der Trinkwasseraufbereitung an den Beispielen Arsenentfernung und zentrale Enthärtung". Dissertation, University Berlin.

Ruijgrok, E.C.M. 2005. "Valuation of nature, water and soil in socio-economic cost benefit. A supplement to the Dutch guideline for CBA." Technical Report.

Rupprecht Consult. 2006. "Evaluation of integrated coastal zone management (ICZM) in Europe". Technical Report, Rupprecht Consult.

RWS, [Rijkswaterstaat]. 1990. "Wassend Water. Gevolgen van het Broeikaseffect voor de Waterstaat." Technical Report.

RWS, [Rijkswaterstaat]. 2005. "EU recommendation concerning the implementation of integrated coastal zone management in Europe. Report on implementation in the Netherlands". Technical Report, Ministry of Transport, Public Works and Water Management, Ministry of Housing, Spatial Planning and the Environment, Ministry of Agriculture, Nature and Food Quality and Ministry of Economic Affairs.

RWS, [Rijkswaterstaat]. 2007a. "ComCoast. The future of flood risk management. A guide to multifunctional coastal defence zones." Final Report, RWS and participating institutions.

RWS, [Rijkswaterstaat]. 2007b. "Workpackage 3: development of alternative overtopping-resistant sea defences phase 3: wave overtopping erosion tests at Groningen Sea Dyke". Final Technical Report, RWS.

RWS, [Rijkswaterstaat]. 2008a. "Golfoverslag en Sterkte Grasbekleding. Uitvoeringsplan 2008". Technical Report, RWS.

RWS, [Rijkswaterstaat]. 2008b. "SBW Belastingen. Projectplan 2008". Technical Report, RWS.

RWS, [Rijkswaterstaat]. 2008c. "SBW Duinen. Projectplan 2008". Technical Report, RWS.

RWS, [Rijkswaterstaat]. 2008d. "SBW Waddenzee. Projectplan 2008". Technical Report, RWS.

RWS Zeeland et al., [Numerous Authors]. 2006. "Gebiedsontwikkleing Perkpolder. Genieten van de elementen. Haalbaarheidsstudie, Juni 2006". Technical Report.

RWS-DWW, [Rijkswatertstaat Dienst Weg- en Waterbouw]. 2006. "The idea behind ComCoast. Innovative solutions for flood protection and regional development". Technical Report.

Salminen, P., J. Hokkanen, and R. Lahdelma. 1998. "Comparing multi criteria methods in the context of environmental problems". *European Journal of Operational Research* 104:485–496.

Schellnhuber, H.-J., and H. Sterr. 1993. *Klimaänderung und Küste. Einblicke ins Treibhaus.* Springer Verlag, Berlin.

Schernewski, G., B. Glaeser, R. Scheibe, A. Sekscinska, and R. Thamm. 2007. "Coastal development: the oder estuary and beyond". Coastline Reports, No. 8, EUCC.

Scholze, M., W. Knorr, N.W. Arnell, and I.C. Prentice. 2006. "A climate-change risk analysis for world ecosystems". *PNAS* 103(35):13116–13120.

Schröter et al., [Numerous Authors]. 2005. "ATEAM Final Report". Technical Report, Potsdam Institute for Climate Impact Research (PIK).

Schuchardt, B., and M. Schirmer. 2005a. Pages 223–239 in "Integrative Analyse und Bewertung der Auswirkungen eines Klimawandels auf die Unterweserregion". In *Klimawandel und Küste. Die Zukunft der Unterweserregion*, 223–239. Springer Verlag, Heidelberg.

Schuchardt, B., and M. Schirmer. 2005b. *Klimawandel und Küste. Die Zukunft der Unterweserregion.* Springer Verlag.

Schuchardt, B., and M. Schirmer. 2007. *Land unter? Klimawandel, Küstenschutz und Risikomanagement in Nordwestdeutschland: die Perspektive 2050.* Oekom Verlag, München.

Schuchardt, B., M. Schirmer, and S. Wittig. 2007. "Klimawandel, Küstenschutz und integriertes Risikomanagement". In *Land unter? Klimawandel, Küstenschutz und Risikomanagement in Nordwestdeutschland: die Perspektive 2050*, 193–215. Oekom Verlag.

Scott, C. 2006. "Wallasea wetland creation scheme – lessons learned". Technical Report, ABPmer Ltd.

Seijffert, J.W. 2001. "Coastal defence in the Netherlands". In *Strategien des Küstenschutzes. Tagungsband zur Veranstaltung am 21. September 2001 in Hannover*, Kolloquien-Reihe. Bilanz und Perspektiven der Umweltpolitik. Niedersächsisches Umweltministerium.

Sindowski, K.-H. 1962. "Nordseevorstöße und Sturmfluten an der ostfriesischen Küste seit 7000 Jahren". *Geographische Rundschau* 14:322–329.

Slager, K. 1992. *De ramp: een reconstructie.* De Koperen Tuin, Goes.

Slocum, N. 2003. "Participatory methods toolkit. A practitioner's manual". Technical Report, King Baudouin Foundation, Flemish Institute for Science and Technology Assessment and United Nations University – Comparative Regional Integration Studies (UNU/CRIS).

Smit, C.J., W.E. van Duin, R.J.H.G. Henkens, and P.A. Slim. 2005. "Casus Hondsbossche Zeewring. Een verkenning van de ecologische effecten van verschillernde kustverdedigingvarianten in de omgeving van de Vereeingde Hardger- en Pettemerpolder". Technical Report, ALTERRA.

Solomon et al., [Numerous Authors]. 2007. "Technical summary". In *Climate Change 2007: The Physical Science Basis. Contribution of Working Group I to the Fourth Assessment Report of the Intergovernmental Panel on Climate Change*, 19–95. Cambridge University Press, Cambridge, UK.

Sorensen, J. 2000. "Baseline 2000 background report: the status of integrated coastal management as an international practice". Technical Report, Harbor and Coastal Center, Urban Harbors Institute, University of Massachusetts.

Spekat, A., W. Enke, and F. Kreienkamp. 2007. "Neuentwicklung von regional hoch aufgelösten Wetterlagen für Deutschland und Bereitstellung regionaler Klimaszena-rios auf der Basis von globalen Klimasimulationen mit dem Regionalisierungsmodell WETTREG". Technical Report, Umweltbundesamt (Ed.).

Stern, N. 2006. "Stern review: the economics of climate change". Technical Report, HM Treasury.

Sterr, H., J.L.A. Hofstede, and H.-P. Plag. 1992. *Proceedings of the International Coastal Congress ICC – Kiel '92*. Dt. Hochschulschriften.

Sterr, H., R.J.T. Klein, and S. Reese. 2000. "Climate change and the coastal zone. An overview of the state-of-the-art on regional and local vulnerability assessment". Working Paper 38.2000, FEEM.

Stock, M. 2002. "Salzwiesenschutz im Wattenmeer". In *Warnsignale aus Nordsee und Wattenmeer. Eine aktuelle Umweltbilanz*, edited by J.L. Lozán, E. Rachor, K. Reise, J. Sündermann and H. von Westernhagen, 364–368. Wissenschaftliche Auswertung, Hamburg.

Stock, M., S. Gettner, H. Hagge, K. Heinzel, J. Kohlus, and H. Stumpe. 2005. "Salzwiesen an der Westküste von Schleswig-Holstein 1988–2001". In *Schriftenreihe des Na-tionalparks Schleswig-Holsteinisches Wattenmeer*, Volume 15. Nationalpark Schleswig-Holsteinisches Wattenmeer.

Stock, M., S. Gettner, J. Kohlus, and H. Stumpe. 2001. "Flächenentwicklung der Festlandssalzwiesen in Schleswig-Holstein". In *Schriftenreihe des Nationalparks Schleswig-Holsteinisches Wattenmeer, Sonderheft*, 57–61. Nationalpark Schleswig-Holsteinisches Wattenmeer, Sonderheft.

Stock, M., K. Kiehl, and H.D. Reinke. 1994. "Salzwiesenschutz im Schleswig-Holsteinischen Wattenmeer: Grundlagen, Zielsetzung und bisherige Umsetzung". Technical Report, UBA-FB 93-101.

Streif, H. 1982. "Geologie des Küstenraumes". In *Das Watt. Ablagerungs- und Lebensraum*, edited by H.-E. Reineck, 24–30. Waldemar Kramer Verlag, Frankfurt.

Streif, H. 2002. "Die Nordsee im Wandel – vom Eiszeitalter bis zur Neuzeit". In *Warnsignale aus Nordsee und Wattenmeer. Eine aktuelle Umweltbilanz*, edited by J.L. Lozán, E. Rachor, K. Reise, J. Sündermann and H. von Westernhagen, 19–28. Wissenschaftliche Auswertung, Hamburg.

Striegnitz, M. 2006. "Conflicts over coastal protection in a National Park: mediation and negotiated law making". *Land Use Policy* 23:26–33.

Stroobandt, T., T. Astley-Reid, N. Houtekamer, and F. Ahlhorn. 2007. "How to write an effective project communication plan?" Technical Report, RWS.

Südbeck, P., and D. Wendt. 2002. "Rote Liste der in Niedersachsen und Bremen gefährdeten Brutvögel, 6. Fassung, Stand 2002". Technical Report.

TAW, [Technical Advisory Committee on Water Defence]. 1998. "Fundamentals on water defence". Technical Report.

TAW, [Technical Advisory Committee on Water Defence]. 2000. "From probability of exceedance to probability of flooding". Technical Report.

Tenge, O. 1898. *Der Jeversche Deichband: Geschichte und Beschreibung der Deiche, Uferwerke und Siele im Dritten Oldenburgischen Deichbande und im Königlichen Preußischen westlichen Jadegebiet*. III. Oldenburgischer Deichband.

Tenge, O. 1912. *Der Butjadinger Deichband: Geschichte und Beschreibung der Deiche, Uferwerke und Siele im Zweiten Oldenburgischen Deichbande und im Königlichen Preußischen östlichen Jadegebiet*. II. Oldenburgischer Deichband.

TERRAMARE. 2001. "Das Ökosystem Wattenmeer. Gesamtsynthese der Ökosystemforschung im Niedersächsischen und Schleswig-Holsteinischen Wattenmeer". Technical Report.

Thomas, K. 2002. "Pre-feasibility study of re-alignment opportunities in Essex prior to implementation of the Essex estuarine strategies and coastal habitat management plan". Technical Report, Environment Agency.

Tomczak, G. 1955. "Was lehrt uns die Holland-Sturmflut 1953?" *Die Küste* 3(1/2):78–95.

UBA, [Umweltbundesamt]. 2007. "Neue Ergebnisse zu regionalen Klimaänderungen. Das statistische Regionalisierungsmodell WETTREG". Hintergrundpapier, Umweltbundesamt (Ed.).

UKCIP, [UK Climate Impacts Programme]. 2003. "Climate adaptation: risk, uncertainty and decision-making". Technical Report Part 2, UKCIP.

UN, [United Nations]. 1992. "Agenda 21 – United Nations Conference on Environment and Development, Rio de Janiero, 3–14 June 1992". Technical Report, United Nations.

UN/ECE, [United Nations Economic Commission for Europe]. 1998. "Convention on access to information, public participation in decision-making and access to justice in environmental matters, Aarhus, Denmark on 25 June 1998". Technical Report.

UNEP-RIVM, [United Nations Environment Programme and National Institute of Public Health and the Environment in The Netherlands]. 2003. "Four scenarios for Europe. Based on UNEP's third global environment outlook". UNEP-DEIA and EW-TR.03-10 and RIVM 402001021, UNEP-RIVM.

UNESCO. 2003. "A reference guide on the use of indicators for integrated coastal management". IOC Manuals and Guides 45, ICAM Dossier.

UVPG. 2005. "Gesetz über die Umweltverträglichkeitsprüfung". BGBl Teil I Nr. 37.

UvW, [Unie van Waterschappen]. 1992. "Water-boards". Brochure, Unie van Waterschappen.

Vagts, I., H. Cordes, G. Weidemann, and D. Mossakowski. 2000. "Auswirkungen von Klimaänderungen auf die biologischen Systeme der Küsten (Salzwiesen & Dünen)". Technical Report.

van de Ven, G.P. 1993. *Man-made lowlands, history of water management and land reclamation in the Netherlands*. ICID and Royal Institute of Engineers in the Netherlands.

van der Linden, H. 1981. "History of the reclamation of the western fenlands and the organisation to keep them drained". In *Proceedings of the symposium on peat lands below sea level*, 42–73. International Institute for Land Reclamation and Improvement/ILRI, Wageningen.

van Duin, W.E., K.S. Dijkema, J. Bunje, T.F. Pedersen, and M. Stock. 1999. "Salt marshes". In *Wadden Sea Quality Status Report 1999*, edited by F. de Jong, J. Bakker, K. van Berkel, N. Dankers, K. Dahl, C. Gätje, H. Marencic and P. Potel, Volume 9, 165–170. Common Wadden Sea Secretariat, Wilhelmshaven.

von Gierke, J. 1917. *Die Geschichte des deutschen Deichrechts. Teil I und II.* Scienta Verlag Aalen.

von Storch, H. 2006. Regionaler Klimawandel in Norddeutschland. Beitrag auf dem Workshop von DKKV und ARL am 27. und 28. November 2006 in Hannover.

von Storch, H. 2007. "Klimaänderungsszenarien". In *Geographie – Physische Geographie und Humangeographie*, edited by H. Gebhardt, R. Glaser, U. Radtke und P. Reuber, 253–256. Spektrum Verlag, Heidelberg.

von Storch, H., and R. Weisse. 2008. "Regional storm climate and related marine hazards in the Northeast Atlantic". In *Climate Extremes and Society*, edited by H.F. Diaz and R.J. Murnane, 54–73. Cambridge University Press, Cambridge, UK.

WBGU, [German Advisory Council on Global Change]. 1999. "Welt im Wandel – Umwelt und Ethik". Sondergutachten.

WBGU, [German Advisory Council on Global Change]. 2000. "World in transition: strategies for managing global environmental risks". Technical Report, WBGU.

WCED, [World Commission on Environment and Development]. 1987. *Our Common Future. World Commission on Environment and Development/WCED (Brundtland's Commission).* Oxford University Press, Oxford, UK.

Weisse, R., and H. Günther. 2007. "Wave climate and long-term changes for the Southern North Sea obtained from a high-resolution hindcast 1958–2002". *Ocean Dynamics* 57:161–172.

Weisse, R., and A. Plüß. 2006. "Storm-related sea level variations along the North Sea coast as simulated by a high-resolution model 1958–2002". *Ocean Dynamics* 56:16–25.

Weisse, R., and W. Rosenthal. 2002. "Szenarien zukünftiger, klimatisch bedingter Entwicklung der Nordsee". In *Warnsignale aus Nordsee und Wattenmeer. Eine aktuelle Umweltbilanz*, edited by J.L. Lozán, E. Rachor, K. Reise, J. Sündermann and H. von Westernhagen, 51–56. Wissenschaftliche Auswertung, Hamburg.

Wilson, S., P. Jones, and W. Sheate. 2003. "Community and public participation: risk communication and improving decision-making in flood and coastal defence". Technical Report, R&D FD2007/TR.

Wittig, S. 2008. "Deichvorländer und Klimawandel: Risikoanalyse und Risikomanagement unter besonderer Berücksichtigung des Küstenschutzes". Dissertation (in preparation), University Bremen.

Wöbcken, C. 1924. *Deiche und Sturmfluten*. Neudruck Walluf bei Wiesbaden.

Wöbcken, C. 1932. *Das Land der Friesen und seine Geschichte*. Schulzesche Verlagsbuchhandlung, Oldenburg.

Wolf, A., and E. Appel-Kummer. 2005. "Demografische Entwicklung und Naturschutz. Perspektiven bis 2015". Technical Report.

Woth, K., R. Weisse, and H. von Storch. 2006. "Climate change and North Sea storm surge extremes: an ensemble study of storm surge extremes expected ion a changed climate projected by four different regional climate models". *Ocean Dynamics* 56:3–15.

WSF, [Wadden Sea Forum]. 2005. "The first steps. Final report". Technical Report.

Zebisch, M., T. Grothmann, D. Schröter, C. Hasse, U. Fritsch, and W. Cramer. 2005. "Klimawandel in Deutschland. Vulnerabilität und Anpassungsstrategien klimasensitiver Systeme". Technical Report 8, Umweltbundesamt (Ed.).

Zimmermann, C., N. von Lieberman, and S. Mai. 2005. "Die Auswirkungen einer Klimaveränderung auf das Küstenschutzsystem an der Unterweser". In *Klima-wandel und Küste. Die Zukunft der Unterweserregion*, 139–166. Springer Verlag, Berlin.

Appendix A
Legal Instruments Related to Multifunctional Coastal Protection Zones

Contents

In this Appendix European and national (especially Lower Saxonian) legal instruments are briefly explained, which have interlinkages to the implementation of Multifunctional Coastal Protection Zones (MCPZ). For example, a detailed discussion concerning the interlinkages between European and national (especially Mecklenburg-Vorpommern) legal instruments can be found in Bosecke (2005).

A.1 Strategic Environmental Assessment (SEA)

The Strategic Environmental Assessment (SEA) was established in 2001 (EC 2001). It deals with the assessment of the effects of plans and programmes on the environment. The objective of the SEA is to support sustainable development and to ensure a high standard for the environment.

In Germany the SEA was implemented by two acts: On the level of town and country planning SEA was implemented by the amendment to the town and country planning code of 20 July 2004 (ICLC 2005). Second, the SEA and the EIA were incorporated in the *Gesetz zur Umweltverträglichkeitsprüfung (UVPG)*. Part I of the UVPG sets the scope of the law, part II covers the EIA and part III covers the prerequisites and the steps necessary if a SEA has to be conducted (UVPG 2005). The annexes of the UVPG stipulates for which plans and programmes a SEA is mandatory and provides a list of criteria on a preliminary case-by-case survey. Paragraph *14o* UVPG stated that for specific water management and spatial planning plans the States has to decide if a SEA is mandatory. On State level it can be chosen between a general adoption of SEA or a special treatment of each

10.1007/978-3-642-01774_BM2,

sector concerned. For Lower Saxony this was applied by the "Niedersächsisches Umweltverträglichkeitsprüfungsgesetz" (NUVPG 2007). For example, in Lower Saxony a SEA was conducted for the amendment of the State Spatial Planning Programme (LROP, ML 2005). According to the criteria in the annex of the NUVPG a SEA has to be carried out for plans and programmes, if they have significant environmental effects (paragraph 9). It is open whether a SEA has to be carried out for the Master Plan for Coastal Protection for Lower Saxony and Bremen published in 2007 (Schuchardt et al. 2007).

A.2 Environmental Impact Assessment (EIA)

The directive on the Environmental Impact Assessment (EIA) was first established in 1985 (EC 1985) and amended in 1997 (EC 1997). The EIA directive forced the Member States to install procedures for the EIA for specific public and private projects. The aim is to assess all possible effects these projects may have on the environment. In article 2 of the directive a duty is placed on Member States to make provisions for the completion of an EIA where projects are "likely to have significant effects on the environment". Annex I and annex II of the EIA (EC 1997) provide lists which are covered by article 2. An EIA is mandatory for the projects listed in annex I, for the projects listed in annex II an EIA is required if the environmental effect is classified as "likely". The projects covered by annex II have to be screened whether an EIA is necessary, in annex III criteria are provided for this decision.

In Germany, the EIA directive has been implemented by the *Gesetz über die Umweltverträglichkeitsprüfung (UVPG)*. Part II of the law covers the EIA purpose, annex I contains a list of projects an EIA is mandatory and annex II provides criteria for a preliminary case-by-case testing. The annexes for mandatory and preliminary testing are slightly different from the EU directive about EIA (IAU 2003). Besides that, the EIA directive has been established in all three Wadden Sea countries, The Netherlands, Denmark and Germany, but with slightly differences. The definition of sensitive areas also differs (in annex III of the EIA directive defined as criterion for preliminary testing: the environmental sensitivity of geographical areas likely to be affected by projects). This may lead to different outcomes whether an EIA is necessary or not. In Lower Saxony, the EIA directive is established as NUVPG (NUVPG 2007). The annex of the NUVPG contains a list of projects and displays for which projects an EIA is mandatory or a preliminary case-by-case testing should be conducted. A preliminary case-by-case testing should be done for the construction of a dam or dike which influences the high water discharge (NUVPG 2007). Additionally, the State Law on Dike of Lower Saxony paragraph 12 no. 1 states that the responsible authority has to conduct a preliminary test whether an EIA is necessary (MU 2004). Furthermore, this paragraph (§12) offers nature conservation organisations the possibility to sue against a planning approval. To enhance the effectiveness the State government released the "10 principles for an effective coastal protection" (MU 2006). Concerning EIA the remarks are as follows:

- An EIA has to be executed if a planning approval is mandatory if significant impacts are likely for the environment. This will be the case for the construction of new dikes. Concerning strengthening and heightening of the main dike §5 point 2 is valid. According to §5 point 2 a measure exceeds the permissibility if it differs considerably from the old dike line. If the case-by-case testing according to the NUVPG states significant environmental effects of the measures, an EIA has to be executed. This is the case if the dike foot is extended seaward. In either case, the impact regulation under nature conservation and the FFH (see Habitat Directive below) compatibility has to be proven.
- For the strengthening and heightening of river dikes no EIA and planning approval have to be conducted, only if the pre-testing according to the NUVPG states no significant effect on the environment.

A.3 Water Framework Directive (WFD)

The Water Framework Directive (WFD, 200/60/EC) was initiated with the purpose of establishing a framework for the protection of inland surface waters, transitional waters, coastal waters and groundwater (article 1, EC 2000). The intention of the framework is further to prevent deterioration and protection and enhancement of the status of aquatic and terrestrial ecosystems and wetlands directly depending on the aquatic ecosystem, promote sustainable water use, ensure the progressive reduction of groundwater pollution and contributes to mitigating the effects of floods and droughts (article 1 paragraph a–e). In article 2 the elements of concern of the WFD are defined. Here, the definition for coastal waters is important (article 2 paragraph 7): *coastal waters* are surface water on the landward side of a line, every point of which is at a distance of one nautical mile on the seaward side from the nearest point of the baseline from which the breadth of territorial waters is measured, extending where appropriate up to the outer limit of transitional waters. The WFD formulates common goals for the planning basis of river basins and necessary measurements to achieve the specific targets. Within the river basin district coastal waters are included and therefore the WFD statements about the status and planning is also valid for the Wadden Sea, e.g. in Lower Saxony. If the concept of a coastal protection zone shall be implemented at the coast, the WFD concerning the status and management of wetlands has to be considered. Within the WFD participation is obliged. To build synergies it is recommended to do the implementation in close cooperation with existing structures e.g. for participation. For more information about the WFD in general see WISE: Water Information System Europe (water.europa.eu), for further information about the application of the WFD in Lower Saxony see website of the NLWKN (www.nlwkn.de).

A.4 Flood Risk Management Directive (FRMD)

The Flood Risk Management Directive (FRMD) proposed by the European Parliament and the European Council is in the consultation phase, the articles of the directive referred to are taken from the proposal of January 18th 2006 (EC 2006). The intention of the FRMD is to enhance the WFD with the aspect of risk at flooding by rivers and adjacent coastal stretches. The FRMD add to the river basin management the aspect of risk of flooding. The FRMD should apply for both rivers and coastal areas. Therefore, the FRMD and its three pillars flood risk assessment, flood risk maps and flood risk management plans are crucial for the concept of coastal protection zones and vice versa. The concept may contribute with solutions to reduce the risk at flooding in coastal areas and provide solutions which should be integrated in flood risk management plans.

Chapter II of the FRMD deals with the preliminary assessment of flood risk (articles 4–6). In article 4 a detailed description of the contents of the flood risk assessment process is given. For example, it shall include a map of the river basin district including the borders of the river basis, sub-basins and where appropriate associated coastal zones showing topography and land use, a description of the flooding process and their sensitivity to change and a description of development plans that would entail a change of land use or of allocation of the population and distribution of economic activities. Article 5 defines the categories to which the river basins, sub-basins or coastal zone should be assigned to: (a) no potential significant flood risk exist or are considered as to be acceptably low or (b) it is concluded that potential significant flood risk exist or might reasonably be considered as likely to occur. There are only two categories of classifying the risk at flooding. The assessment has to be done at latest three years after the date of entry into force, and the review shall be done at latest until 2018 and every six years thereafter.

Chapter III deals with the preparation of flood risk maps (article 7–8). In article 7 detailed information is given about the categories which should be covered by the flood risk maps: (a) floods with high probability (once every ten years), (b) floods with medium probability (once every hundred years) and (c) floods with low probability (extreme events). For these categories the following information should be given: projected water depth, the flow velocity, the areas which could be subject to bank erosion and debris flow deposition. The third part of article 7 deals with indicative flood damage maps which should show: (a) the number of inhabitants potentially affected, (b) potential economic damage in the area and (c) potential damage to the environment. The first aspect is relatively easy to show, but the last two aspects are rather difficult to process in maps. Many research projects dealt with the determination of potential economic damage after floods, and all of them concluded that the values are estimations or tendencies. In Germany, one research project was conducted on the micro scale level to investigate the potential of flood damage: Micro Scale Evaluation of Risks in Flood prone Coastal Zones (MERK, Reese et al. 2002). Article 8 formulates that the flood risk maps should be completed until 2013 at latest and should be updated every six years thereafter.

Chapter IV comprises the flood risk management plans, the third pillar of FRMD (article 9–12). These articles deals with the development and implementation of flood risk management plans in vulnerable river basins and coastal areas and the coordination mechanisms of the management plans within the river basin districts (EC 2006, p. 8). An important aspect is mentioned in article 9 paragraph 1: the Member States shall prepare and implement flood risk management plans at the level of the river basin district for the river basins, sub-basins and stretches of coastline identified under point (b) of article 5 paragraph 1. Another important aspect is mentioned in paragraph 2: Member States shall establish appropriate levels of protection specific to each river basin, sub-basin or stretch of coastline, focusing on the reduction of the probability of flooding and of potential consequences of flooding to human health, the environment and economic activity, and taking into account relevant aspects such as water management, soil management, spatial planning, land use and nature conservation.

A.5 Birds Directive

"The Birds Directive 79/409/EEC (EC 1979) was the first piece of EU nature conservation legislation. Under the Directive Member States are under duty to take measures to maintain a sufficient diversity of habitats for all European wild birds and regularly occurring migratory birds. The duty extends to the creation of Special Protection Areas (SPA). Once a SPA has been designated, the Member States must take steps to avoid deterioration of the habitat, or pollution or the disturbance of the birds within it. A second part of the Directive relates to a number of bans on activities that directly threaten birds and associated activities such as trading in live or dead birds. A further component of the Directive establishes rules that limit the number of species that can be hunted and the periods during which they can be hunted. There are procedures within the Directive that allow for the granting of consents to authorise activities that would be harmful to habitats and species" (IAU 2003, p. 26). Designated under the Birds Directive are almost all areas of the Wadden Sea National Park in Lower Saxony, except three areas and the recreation zone above mean high water level.

A.6 Habitats Directive

"The Habitats Directive 92/43/EEC (EC 1992) on the Conservation of Natural Habitats and of Wild Fauna and Flora provides for the creation of an Europe wide network of Special Areas of Conservation (SAC), known collectively as NATURA 2000. This is to be a coherent ecological network consisting of the sites that meet the criteria provided in Annex I of the Directive and those sites designated as SPAs under the Birds Directive" (IAU 2003, p. 27). "The second important feature of the Habitats Directive is the introduction, under Article 6 (3) and 6 (4), of a formal procedure for assessing whether projects or plans, either alone or in combination with other projects or plans, are likely to have a significant effect on a NATURA 2000 site. Where significant effects are envisaged, an 'appropriate assessment' of the project or plan must be completed. This assessment is stage by stage consideration of key factors" (IAU 2003, p. 27). Coastal protection measures have to consider this assessment procedure if they have a significant effect on designated habitat sites (NLWKN 2007b). Consequently, this is also valid for the implementation of the concept of coastal protection zones, because parts of the Wadden Sea National Park are submitted to be designated under the Habitat Directive: the areas will cover the restricted zone and intermediate zone.

A.7 Law on Water Management of Lower Saxony (NWG)

The purpose of the Law on Water Management of Lower Saxony (Niedersächsisches Wassergesetz, NWG) are surface water, coastal waters and ground water (NWG 2004). Coastal waters are defined as "sea between the coastline during mean high water or the seaward border of surface water and the seaward border of the territorial sea" (Küstenmeer) (article 1 (1)). The coast line during mean high water is the coast of Lower Saxony defined as water level of mean high tide water (article 1 (4)). In article 1 (5) according to the WFD the NWG defines river basins, sub-basins and river basin districts.

In article 2 the general principles of the treatment of water bodies and the water management are formulated. Article 2 (1) describes the water bodies as part of the natural environment and habitat for plants and animals which have to be adequately protected. Article 2 (2) formulates specific requirements such as the prevention of high water and the wash away of soil or that the water bodies are relevant for plants and animals and that the relevance for the landscape should be taken into account. Article 3 defines the river basin districts of the river Ems, Weser and Elbe for Lower Saxony. The river basin district of the Weser for example comprises the Wadden Sea from the east border of the river basin district of the Ems until the river basin district of the Elbe which is the main part of the Wadden Sea in Lower Saxony. The NWG with its latest amendment in 2007 should apply the European Directives for the EIA and the WFD. In the third part of NWG rules for the treatment of coastal waters are established. Article 132 deals with the reconstruction of coastal waters. Article 132 (1) states that for the reconstruction of coastal waters a plan approval is mandatory. Even for the construction, deconstruction or significant changes of coastal protection elements a plan approval is mandatory. Paragraph 2 formulates that a plan approval is redundant if a scheme according article 132 (1) does not need an EIA.

Appendix B
European Case Studies

Contents

This Appendix comprises three European case studies of the ComCoast project which exemplify both the performance and the results of the participation processes.

B.1 Case Study Abbotts Hall (UK)

B.1.1 Description of the Setting and the Process

The farm of Abbots Hall lies at the Blackwater Estuary in East Anglia. The Blackwater Estuary is the largest estuary in East Anglia covering almost 4,400 ha and it is an internationally important area for wildlife. About 40% of the salt marshes in Essex were lost due to erosion (*coastal squeeze*) over the last 25 years. Abbotts Hall farm is located 10 km south of Colchester and covers 287 ha farm land. Different organisations (Essex Wildlife Trust, English Nature, WWF-UK, and the Environment Agency, EA) jointly develop almost 80 ha into wetland for a sustainable flood defence. This was necessary because the condition of the sea wall was too bad to ensure the use of the farm land. The obligatory cost-benefit analysis shows that maintenance work for the sea wall will be to expensive to protect the farm land. Thus, alternatives have to be found. The Essex Wildlife Trust has purchased Abbotts Hall farm in 2000. The planning for the sustainable flood defence development took 2 years for getting planning and other permissions. In autumn 2002 four breaches were made in the existing embankment, a fifth was added half a year later (see

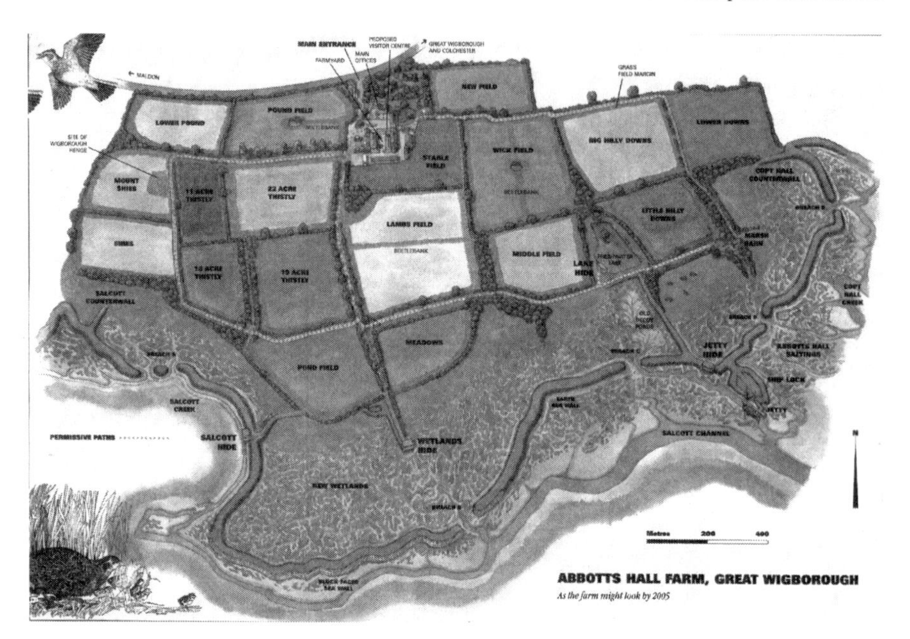

Fig. B.1 Plan of the Abbotts Hall farm with the breaches and the new created salt marsh (see *bottom*) and the farm house and the proposed information center (see *top*)
Source: (EWT 2004).

Fig. B.1). In 2005 the Chartered Institution of Water and Environment Management (CIWEM) and the Royal Society for the Protection of Birds (RSBP) awarded Abbotts Hall farm with their "Living Wetland Award" (EA 2005b).

Local knowledge of project-partners was the basis for a comprehensive overview on relevant stakeholders in the area of concern. Additionally, an extended stakeholder analysis was made via postcodes. Due to the likely negative influence to nature an Environmental Impact Assessment (EIA) was carried out and used as a formal instrument for consultation and communication with the public. Within this EIA several meetings and public talks were held. Furthermore, the partners executed several meetings targeting specific interests, e.g. the oystermen in the proximity worried about the opening of the dike because of the prospect of increased organic matter in the area. Most of the concerns were addressed and resolved during the EIA, but some of them were left open. These concerns were met by direct consultation and one-to-one meetings. For example, the decline of oyster stocks were related to an adjacent sewage treatment work by conducting random water quality monitoring (EA 2005b).

B.1.2 Results of the Participation Process in Abbotts Hall

One main result were the acceptance of the breaches in the embankment: In 2002 the breaching was accompanied by a big event of approx. 2,000 people. This was a result of the long process achieved by formal and informal participation procedures.

Abbotts Hall farm is now converted from an old farm house to a farm with different kinds of possibilities. For example, the cropping regime on the arable farm has been changed to increase the diversity of crops, some hectares of arable fields have been changed into organic production and sheep are grazing on Abbotts Hall again (EWT 2004). In the first year after the breaches up to 1,700 brent gees have been counted in the salty area and hundreds of waders came and roost and feed in the area. The first fish survey shows that about 10 species were using the area as feeding and breeding ground. "After the autumn breaching of the sea wall and the high spring tides that followed, seeds of salt marsh plants flooded into the site. To our great delight and, it has to be said, considerable relief, the first shoots of salt marsh plants emerged in the spring. The plants that have colonised include glassworts (Samphire), grass leaved orache (Atriplex species), sea spurry (Spergularia rubra) and annual sea blite (Suaeda maritima). This plant community is characteristic of open situations free of competition from established perennials, which will develop later" (EWT 2004, p. 2). One major contribution of the ComCoast project was the support in implementing the proposed information centre. This centre will present and provide information on the implemented scheme, the monitoring results and on the interlinkages between nature, climate and the coast.

The implementation process for the Abbotts Hall project has several positive side effects. The leading agency for the process was the Environment Agency (EA); it was recognised after this process as a reliable partner throughout all stakeholders with staff that listen and taking the raised issues into account. Furthermore, the process has built strong relationships between the EA and stakeholders, which are now in the position of working together and to determine joint goals and approaches. Before this process was started, there was no formal procedure to run a participation process, hence, this approach of stakeholder engagement was developed during the first project phase. "The experience has revealed the importance of interpersonal skills... in establishing good lines of communication with stakeholders" (EA 2005b, p. 13). Lessons learned from the stakeholder engagement process at Abbotts Hall were, that the following questions have to be answered before conducting a participation process:

- Who are the main stakeholders?
- Are the expectations clear?
- Are there real benefits from participating?
- Is the team working?

One success factor is the tailor-made application of specific approaches and methods during the process ranging from "general meeting" targeting only the provision of information to "one-to-one meeting" discussing key issues with stakeholders.

B.2 Case Study Wallasea Wetland Creation (UK)

The port development projects along the Essex coastline led to loss and destruction of salt marshes and mudflats which have to be compensated. The area needed to compensate the loss was 108 ha, the destroyed habitat was 54 ha. Because of the delay of the compensation and additional items it was agreed to compensate "two-for-one". An additional aim of the compensation was to create new habitat which is big enough to be accepted by seabirds. Another added-value was the enhancement of the coastal protection of Wallasea Island which was at risk of flooding. The port development projects were undertaken in 1988 and 1994 and a legal action of The Royal Society for the Protection of Birds (RSPB) led to a judgement by the European Court of Justice in 1996 and further threats if compensation was not applied. The planning and building phase was relatively short, only 30 months, but the site selection phase took about 7 years (Scott 2006).

The following sections summarise the methods of identifying Wallasea Island as appropriate site to compensate habitat loss of salt marshes and mudflats.

B.2.1 Steps to Identify a Feasible Site

In 1997 Halcrow was commissioned to conduct an investigation along the Essex coastline for presenting a coherent management strategy for the Essex tidal defences. The protected area is about 15,600 ha with high property and agricultural asset values (Halcrow 1998a). The proposed strategies in the Halcrow report were divided into:

1. Improve to a standard recommend by recent project appraisal or under construction.
2. Maintain the short term, but consider managed retreat/setback in the long term only if the resulting geomorphological change is shown to produce a more sustainable estuary.
3. Undertake detailed project appraisal and if justified carry out improvements.
4. Maintain the short and long term (Halcrow 1998a, p. 2).

"The proposed strategies comply with the findings of the SMP (Shoreline Management Plan). One of the key aspects of the SMP is that the estuaries should be hydrodynamically and geomorphologically modeled to assess the effects of the sea level rise and possible managed retreat. In the meantime it recommends a policy of holding the existing line until the results of such a model are available. Even when the results can be reviewed, any retreat should only be implemented where setback can be shown to have a positive geomorphological effect on the sustainability of the estuary" (Halcrow 1998a, p. 3). The study identifies about 111 km which can be taken under strategy 2 with different levels of appropriateness.

Before the Essex Sea Wall Management strategy was carried out, a pilot study was performed to validate and improve the feasibility of the approach, e.g. to prove the quality and availability of data. The management strategy has considered different aspects (Halcrow 1998a, p. 4, for further details see Halcrow 1998b):

- the economic viability of the existing defence,
- environmental considerations, the SMP and other available documents,
- present and future modeling capacity,
- programming of capital works,
- future works and studies.

In 2002 the Environment Agency performed a pre-feasibility study to identify sites for managed realignment in the estuaries of Essex (Thomas 2002). On the basis of the Halcrow survey the following steps were taken:

- LIDAR: Using **L**aser **I**lluminated **D**etection and **R**anging (LIDAR) data to easily identify whether the site has appropriate elevation or not
- Discussion: Discussion between local staff using different sources of data and information to pre-select feasible sites
- Scoring: A series of scoring criteria were agreed prior to the discussion with advice from Halcrow
- Weighting: The scores need to be weighted, as some criteria are considered to be more important than others

In Fig. B.2 the categories and the assessment criteria which were applied in classifying the feasible managed realignment sites are shown. Wallasea Island gets 415 weighted points. Finally, Wallasea Island was ranked on the second position of all managed realignment sites in Essex.

Category	Assessment Criteria	Possible Points (*min threshold level)	Weighting
Costs	*Cost per hectare of habitat created* Ranging from minimal work to maximum work	5,4,3,2,1	*20
Hydro-dynamics	*Potential for impact on coastal/estuarine processes* Ranging from minimum impact to maximum impact likely.	5,4,3*,1,0	*25
Environment	*Potential to improve the local environment* Ranging from positive improvement to neglible improvement.	5,4,3*,1,0	*15
Owner Interest	*Is the land owner interested in managed realignment on his/her land?* Ranging from all landowners are interested to no landowner interested/Landowner view not known.	5,3,0	*20
Defence Condition	*What is the standard of the current flood defence at this location?* Ranging from excellent to very poor.	1,2,3,4,5	*20

Fig. B.2 Listing of categories and assessment criteria to identify managed realignment sites Source: Thomas (2002).

The final process of identifying and choosing Wallasea for management realignment was done by a combination of multi-criteria analysis and consultation processes. Within the multi-criteria analysis hydrodynamical and ecological aspects have been considered. The Department for Environment, Food and Rural Affairs (DEFRA) has conducted a public consultation process to gather the views of the local community and selected authorities. For Weymarks (another designated compensation site in Essex) a comprehensive consultation process was undertaken, but due to additional benefits Wallasea was chosen as preferred option. The special situation in Wallasea was, that the landowner had already recognised that it was not economic feasible to protect the land on the north bank through the enhancement of the existing defence line. Therefore, the landowner had built a new embankment in the hinterland. So, there was a big chance to use the land in front of the new dike for compensatory habitat.

B.2.2 Wallasea Island Today

> On the 4th July 2006 sea wall breaching works was completed on the Wallasea Managed Realignment Scheme in the Crouch Estuary (Essex). The site now represents one of the largest coastal realignment area in Europe; it is 4 km length, 108 ha in extent and, on each tide betwee 790,000 m^3 and 1.7 million m^3 of water (neap and spring tides respectively) are exchanged with the adjacent estuary (Scott 2006, p. 1, see Fig B.3).

Fig. B.3 Aerial photograph of Wallasea wetland
Source: www.abpmer.net/wallasea

B.3 Case Study Perkpolder (NL)

B.3.1 Description of the Setting and the Process

The pilot area Perkpolder is located in the southern part of the Province of Zeeland (see Fig. B.4). A former car ferry harbour (out of duty since March 2003 due to a tunnel under the Westerschelde) and an adjacent polder are proposed for compensation of estuarine wetland compensation by the Ministry of Transport, Public Works and Water Management and the Province of Zeeland. This habitat is expected to be lost due to the already carried out dredging of the Westerschelde for container shipping to Antwerp. Parallel, the municipality of Hulst has developed a vision for the area to give an impulse to economic activities. On the other hand, there was the demand from the European Union to apply the compensation of habitat before the next dredging will take place at the Westerschelde (RWS Zeeland et al. 2006).

The vision of the municipality comprises a marina in a part of the ferry harbour, a golf course, water accommodated living and upgrading of the landscape and recreational activities. The water board wants to improve an inner embankment due to coastal protection reasons. In 2005 the municipality of Hulst has searched for private project developers which should develop the area taking the vision into account. Two private companies were selected. The consortium of municipality and private companies carried out a feasibility study for the area of concern. The feasibility study contains a market analysis, a Strength-Weakness-Option-Threats (SWOT)-analysis and an analysis of possible target groups.

First, the participation process was executed with selected stakeholders (RWS Zeeland et al. 2006). But, a comprehensive stakeholder analysis revealed that not all relevant parties were involved. These parties were subsequently involved. At this

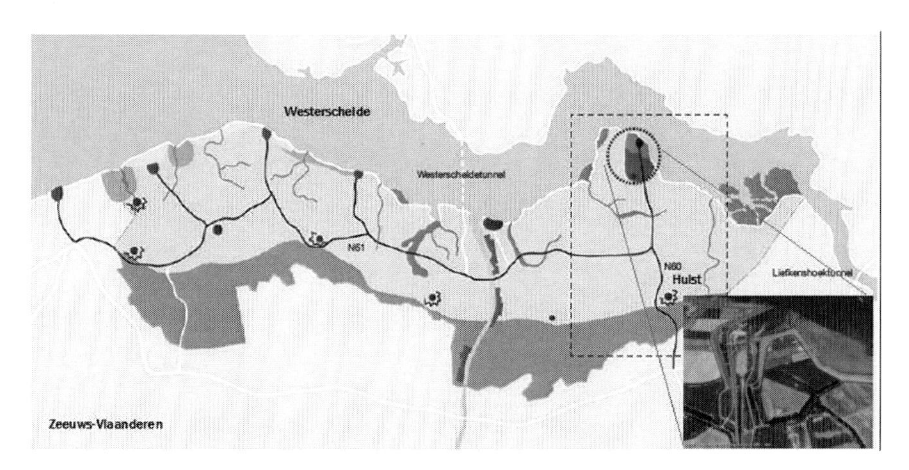

Fig. B.4 The region Zeeuws-Vlaanderen in the southern part of The Netherlands adjacent to Belgium. The aerial photograph shows the former harbour area at Perkpolder
Source: RWS Zeeland et al. (2006).

stage, the process was attended only by organisations and institutions. From the end of 2005 to the beginning of 2006, several workshops were carried out, targeting specific issues with special working groups if necessary. The results produced by the working groups were presented and jointly discussed by the consortium. The broad public has been informed about the plans in the course of the year 2006.

B.3.2 Results of the Participation Process

The Perkpolder plan is not realised yet. But, several results of the participation process can be presented. One result belongs to the ComCoast project, which had a major influence on the development of the old ferry harbour area into a multifunctional coastal protection zone with the implementation of overtopping resistant elements. The idea of the ComCoast project also stimulated to combine the obliged compensation for dredging and the development of the area (RWS Zeeland et al. 2006). Figure B.5 shows the present plan for the area. The former harbour area will be transferred into a marina with holiday houses in the proximity. The area behind the main dike on the eastern part will be developed to a salt marsh with tidal influence (compensation area). The main dike will be breached. The dike will also be used for recreational activities like bird watching and walking. The tides should also influence the western part of the area. Therefore, the road will be tunneled by a sluice. The whole area will be developed under the circumstances that the embankments have to be adapted to a sea level rise in the future. The construction of the main embankment as overtopping resistant (a result of the ComCoast intervention) and the adaptation of the water management system to these requirements offers further opportunities in the future. Currently, it is planned to create a tidal golf course in that area right behind the dike (see Fig. B.5b).

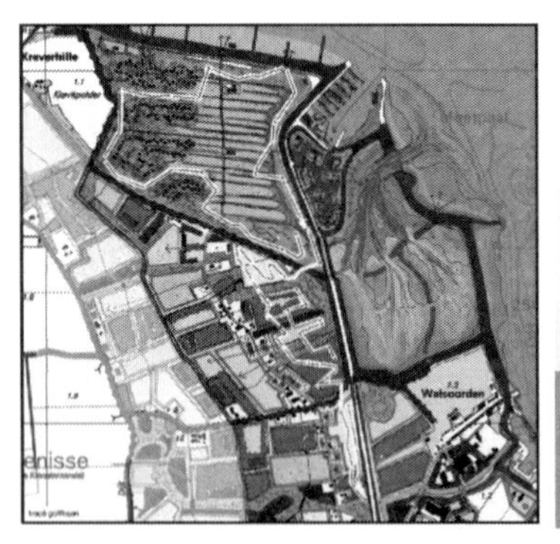

(a) Plan for the Old Ferry Harbour

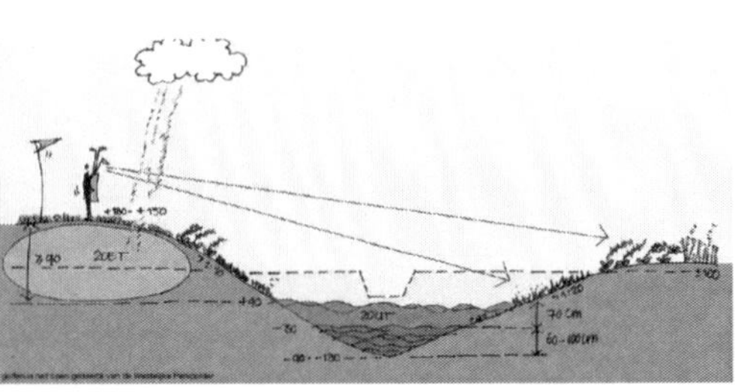

(b) Drawing of the Tidal Golf Course

Fig. B.5 The proposed plan for the pilot area Perkpolder in the Province of Zeeland (**a**) and an example for the tidal golf course in the transition zone (**b**)
RWS Zeeland et al. (2006).

Index